Urban Transportation and Logistics

Health, Safety, and Security Concerns

Urban Transportation and Logistics

Health, Safety, and Security Concerns

Edited by
Eiichi Taniguchi
Tien Fang Fwa
Russell G. Thompson

CRC Press
Taylor & Francis Group
Boca Raton   London   New York

CRC Press is an imprint of the
Taylor & Francis Group, an **informa** business

CRC Press
Taylor & Francis Group
6000 Broken Sound Parkway NW, Suite 300
Boca Raton, FL 33487-2742

© 2014 by Taylor & Francis Group, LLC
CRC Press is an imprint of Taylor & Francis Group, an Informa business

No claim to original U.S. Government works

Version Date: 20131112

International Standard Book Number-13: 978-1-4822-0909-9 (Hardback)

This book contains information obtained from authentic and highly regarded sources. Reasonable efforts have been made to publish reliable data and information, but the author and publisher cannot assume responsibility for the validity of all materials or the consequences of their use. The authors and publishers have attempted to trace the copyright holders of all material reproduced in this publication and apologize to copyright holders if permission to publish in this form has not been obtained. If any copyright material has not been acknowledged please write and let us know so we may rectify in any future reprint.

Except as permitted under U.S. Copyright Law, no part of this book may be reprinted, reproduced, transmitted, or utilized in any form by any electronic, mechanical, or other means, now known or hereafter invented, including photocopying, microfilming, and recording, or in any information storage or retrieval system, without written permission from the publishers.

For permission to photocopy or use material electronically from this work, please access www.copyright.com (http://www.copyright.com/) or contact the Copyright Clearance Center, Inc. (CCC), 222 Rosewood Drive, Danvers, MA 01923, 978-750-8400. CCC is a not-for-profit organization that provides licenses and registration for a variety of users. For organizations that have been granted a photocopy license by the CCC, a separate system of payment has been arranged.

Trademark Notice: Product or corporate names may be trademarks or registered trademarks, and are used only for identification and explanation without intent to infringe.

Visit the Taylor & Francis Web site at
http://www.taylorandfrancis.com

and the CRC Press Web site at
http://www.crcpress.com

Contents

	Preface	vii
	The Editors	ix
	Contributors	xi
Chapter 1	Concepts and Visions for Urban Transport and Logistics Relating to Human Security *Eiichi Taniguchi, Russell G. Thompson, and Tadashi Yamada*	1
Chapter 2	Transport and Logistics in Asian Cities *Tien Fang Fwa*	31
Chapter 3	Healthy Transport *Russell G. Thompson*	53
Chapter 4	Hazardous Material Transportation *Wai Yuen Szeto and Rojee Pradhananga*	77
Chapter 5	Mixed Traffic in Asian Cities *Yasuhiro Shiomi*	101
Chapter 6	Road Safety *Nobuhiro Uno, Yasunobu Oshima, and Russell G. Thompson*	123
Chapter 7	Network Design for Freight Transport and Supply Chain *Tadashi Yamada*	167
Chapter 8	Vehicle Routing and Scheduling with Uncertainty *Ali Gul Qureshi*	189

Chapter 9	Urban Transport and Logistics in Cases of Natural Disasters *Sideney A. Schreiner, Jr.*	225
Chapter 10	Application of ICT and ITS *Takayoshi Yokota and Dai Tamagawa*	245
Chapter 11	Future Perspectives on Urban Freight Transport *Eiichi Taniguchi and Russell G. Thompson*	255
Index		261

Preface

Urban transport and logistics systems play a very important role in human security engineering since they provide basic components for ensuring the safety and security of human life in urban areas. Most of the activities of people in business and leisure depend greatly on the services of urban transport and logistics systems. However, we face difficult and complicated problems of efficiency, environment, energy consumption, safety relating to urban transport, and logistics in normal cases as well as in disasters. Therefore, understanding the problems, finding approaches and solutions, implementing them, and evaluating results are essential for creating better urban planning and policy implementation.

This textbook aims to provide advanced knowledge and experience on urban transport and logistics for human security engineering. It includes a wide range of subjects:

- Concepts and vision for urban transport and logistics relating to human security
- Transport and logistics in Asian cities
- Healthy transport
- Hazardous material transport
- Mixed traffic in Asian cities
- Road safety
- Network design for freight transport and supply chain
- Vehicle routing and scheduling with uncertainty
- Urban transport and logistics in natural disasters
- Application of ICT (information and communication technology) and ITS (intelligent transport systems)
- Future perspectives on urban freight transport

These subjects provide an important basis for discussing transport and logistics systems in urban areas from viewpoints of safety and security of human life. The ideas and knowledge included in these subject areas are relatively new, but are very useful for creating innovative solutions to tackle real problems in urban areas.

This textbook was mainly prepared for the urban transport and logistics course in the graduate school of the Global Center of Excellence program, "Human Security Engineering for Asian Megacities," in which Kyoto University, the National University of Singapore, and Monash University jointly participated. Researchers of these three universities discussed and lectured students in the graduate schools. However, it will also be useful for practitioners who are involved in urban transport and logistics planning, since it presents examples and case studies in real situations. We hope that this textbook will be used for studying human security engineering as well as developing a higher quality of life in urban areas.

The Editors

Eiichi Taniguchi is professor of transport and logistics in the Department of Urban Management, Graduate School of Engineering, Kyoto University, Japan. His research centers on city logistics and urban freight transport modeling focusing on stochastic and dynamic vehicle routing and scheduling with time windows, multi-agent and simulation considering behavior of stakeholders who are involved in urban freight transport. Currently, his research covers the health and security issues including humanitarian logistics after catastrophic disasters and home health care problems in aging society. He has published more than 200 academic papers and nine books. He received the best paper award from the Japan Society of Civil Engineers in 2000 as well as from the Eastern Asia Society for Transportation Studies in 1999 and 2011. He has organized the First to Eighth International Conferences on City Logistics in various venues in the world as the president of the Institute for City Logistics since 1999. He has been actively involved in collaborative research in international organizations including Organisation for Economic Co-operation and Development, World Conference on Transport Research Society, Transportation Research Board, and World Road Association.

Dr. T. F. Fwa is professor in the Department of Civil Engineering and director of the Centre for Transportation Research, National University of Singapore. He received his BEng (First Class Hons.) from the then University of Singapore (now known as the National University of Singapore), his MEng from the University of Waterloo, Canada, and his PhD from Purdue University, USA.

Dr. Fwa is active academically and professionally in the area of transportation infrastructure engineering. He has been invited to lecture and make technical presentations in sixteen countries, including keynote lectures at a number of international conferences and symposia. He has been either the principal or co-investigator for twenty-one funded research projects since 1985. He has published more than 200 technical papers in international journals and conferences. He has received a number of awards for his academic and research contributions. The awards he received include the 1992 Arthur M. Wellington Prize from the American Society of Civil Engineers; the 2000 Engineering Achievement Award from the Institution of Engineers, Singapore; the 2005 Frank M. Masters Transportation Engineering Award from the American Society of Civil Engineers, USA; and the 2009 Alfred Noble Prize from the American Society of Civil Engineers, USA.

Dr. Russell G. Thompson, BAppSc, MEngSc, PhD, is currently the director of the Master of Transportation Systems program at the Joint Southeast University and Monash University Graduate School in Suzhou, China. He holds a bachelor's degree in mathematics, master's degree in traffic and transport engineering, and a PhD in traffic modeling.

Dr. Thompson was a founding director and has been vice president of the Institute of City Logistics, based in Kyoto, since 1999. He is a core research partner in the Center of Sustainable Urban Freight Systems, a Volvo Center of Excellence. Dr. Thompson has coauthored over 160 research publications, including over 70 refereed publications. He has also coauthored three research books as well as seven international conference books.

Dr. Thompson has been a member of Kyoto University's Global Center of Excellence in Human Security Engineering since 2008. He is currently a partner researcher in the Concert Japan Program on Resilience against Disasters and a lead researcher on the project Improving the Resilience of Road Freight Networks under Earthquakes. Since 2012, Dr. Thompson has been a senior research fellow in humanitarian logistics at the Australian Defense Force Academy. He has been involved in a number of research projects associated with infrastructure recovery and humanitarian logistics after the Tohoku disasters in Japan in 2011.

Contributors

Tien Fang Fwa
National University of Singapore
Singapore

Yasunobu Oshima
Kyoto University
Kyoto, Japan

Rojee Pradhananga
Kyoto University
Kyoto, Japan

Ali Gul Qureshi
Kyoto University
Kyoto, Japan

Sideney A. Schreiner, Jr.
EWS Engenharia de Transportes
 Ltda
São Paulo, Brazil

Yasuhiro Shiomi
Ritsumeikan University
Kyoto, Japan

Wai Yuen Szeto
The University of Hong Kong
Hong Kong, China

Dai Tamagawa
Hanshin Expressway R&D
 Company Limited
Osaka, Japan

Eiichi Taniguchi
Kyoto University
Kyoto, Japan

Russell G. Thompson
Monash University
Melbourne, Australia

Nobuhiro Uno
Kyoto University
Kyoto, Japan

Tadashi Yamada
Kyoto University
Kyoto, Japan

Takayoshi Yokota
Tottori University
Tottori City, Japan

CHAPTER 1

Concepts and Visions for Urban Transport and Logistics Relating to Human Security

Eiichi Taniguchi, Russell G. Thompson, and Tadashi Yamada

CONTENTS

1.1	Concepts of City Logistics	2
1.2	Visions for City Logistics	3
	1.2.1 General Remarks	3
	1.2.2 ICT, ITS, and City Logistics	5
	1.2.3 City Planning and City Logistics	5
	1.2.4 Land Use Planning and City Logistics	6
	1.2.5 Units of Urban Freight Transport Planning	6
	1.2.6 Subsidies and Additional Charges from the Public	6
1.3	Incorporating Risks in Urban Freight Transport	7
1.4	Methodology	10
	1.4.1 Robustness	10
	1.4.2 Stochastic Programming	11
	1.4.3 Simulation	12
	1.4.4 Multiobjective Optimization	12
	1.4.5 Multiagent Simulation	14
1.5	Health	15
	1.5.1 Introduction	15
	1.5.2 Air Quality	16
	1.5.3 Physical Activity	17
1.6	Human Security Engineering	18
	1.6.1 Introduction	18
	1.6.2 Man-Made Disasters	19

1.6.3 Hazardous Material Transport   20
1.6.4 Traffic Safety   21
1.7 Conclusion   21
References   22

1.1 CONCEPTS OF CITY LOGISTICS

Urban transport and logistics issues present many difficult problems. There are four major stakeholders—shippers, freight carriers, residents, and administrators—who have different goals and are involved in various initiatives. On one hand, we need to take an approach based on the industrial point of view, which allows us to establish more efficient and competitive logistics systems for supporting "just-in-time" production and delivery systems. Shippers in general hope to receive and send their goods in a reliable manner that does not violate the designated time window of delivery to lower their delivery costs. Freight carriers try to meet the shippers' needs using their resources and public infrastructure and information to maximize their profits. On the other hand, residents in urban areas require minimum nuisance from urban freight transport and desire safer and more comfortable communities. Administrators of municipalities try to enhance the quality of life for residents as well as decrease the congestion levels within the urban road network, decrease the negative environmental impacts, and increase security relating to urban freight transport.

To address these complicated and difficult problems, the concept of city logistics was proposed. Taniguchi, Thompson, Yamada, and van Duin (2001) defined city logistics as the following:

> City logistics is the process for totally optimizing the logistics and transport activities by private companies with support of advanced information systems in urban areas considering the traffic environment, the traffic congestion, the traffic safety and the energy savings within the framework of a market economy.

City logistics promotes the establishment of efficient, safe, and environmentally friendly urban logistics systems using innovative technologies of ICT (information and communication technology) and ITS (intelligent transport systems) as well as the environmental appraisal. These two driving forces of innovative technology and environmental appraisal facilitate the planning and implementation of city logistics measures for sustainable development of urban areas.

1.2 VISIONS FOR CITY LOGISTICS

1.2.1 General Remarks

Why do we need visions? We have already defined city logistics in the previous section. That statement provides us a conceptual idea of what city logistics are. However, in order to establish efficient, safe, and environmentally friendly urban logistics systems through the process of city logistics, we need visions for city logistics.

First of all, it is necessary to set targets for the activities that can be achieved using city logistics. In this context we would like to consider four pillars (as shown in Figure 1.1):

a. Mobility
b. Sustainability
c. Livability
d. Resilence

Mobility is a basic requirement for transporting goods within as well as into and from urban areas. Reliable road, rail, and other modal

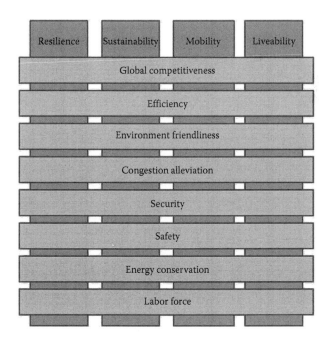

FIGURE 1.1 Structure of visions for city logistics.

networks are essential in terms of connectivity and travel times. Providing enough road network capacity and alleviating traffic congestion are always important in the agenda of urban traffic management. In particular, it is vital for urban freight transport, since many freight carriers have to meet severe time windows set by customers within the framework of just-in-time transport systems.

Sustainability has become more important, since people are concerned about environmental issues including air pollution, noise, vibration, and visual intrusion. Large freight vehicles are often the source of these negative environmental effects. Therefore, minimizing the negative impacts on the environment by trucks is an important issue to be addressed when managing urban freight transport systems. As well, minimizing energy consumption is required to ensure a sustainable city.

Livability should be taken into account when planning urban logistics systems. Residents in urban areas enjoy the benefits of buying a wide variety of commodities based on urban delivery systems to retail shops or even directly to homes. But residents are also concerned about traffic safety and environment in local areas, which may be threatened by heavy commercial vehicles traveling within and near residential areas.

Recently, resilience has become more important due to the increasing threats of natural and man-made hazards. Infrastructure in urban areas should be improved to mitigate the damage caused by natural and man-made hazards to make a quick recovery after such events. City logistics can contribute to increasing the level of urban resilience and provide a safer and higher quality of life.

Therefore, the visions of city logistics are to create mobile, sustainable, and livable cities by supplying the necessary goods for activities and collecting goods that are produced in the city, as well as minimizing negative impacts on the environment, safety, and energy consumption.

As Figure 1.1 indicates, mobility, sustainability, and livability, as well as resilience, are four pillars of the visions for city logistics. They represent the goals of city logistics and are supported by societal values that brace the structure of the visions, comprising:

    a. Global competitiveness
    b. Efficiency
    c. Environmental friendliness
    d. Congestion alleviation
    e. Security
    f. Safety
    g. Energy conservation
    h. Labor force

These values must be considered in city logistics to achieve more mobile, sustainable, and livable cities. Urban freight problems are varied and complex since they are associated with economic, financial, environmental, and social aspects.

1.2.2 ICT, ITS, and City Logistics

In the twentieth century there was considerable discussion that focused on the trade-off relationship between efficiency and environmental friendliness of urban freight transport systems. That is, if freight carriers try to establish more efficient urban freight transport systems, they will generate more negative impacts on the environment. In contrast, if they try to use environmentally friendly urban delivery systems, they have to pay extra costs for lowering hazardous gas emissions and noise (e.g., Cooper 1991). However, in the twenty-first century we are in a good position of being able to use information communication technology (ICT) and intelligent transport systems (ITS) for overcoming the trade-off relationship between efficiency and environmental friendliness.

ICT and ITS enable efficient and environmentally friendly urban delivery systems to be developed. An example of this is ITS-based probabilistic or dynamic vehicle routing and scheduling planning with time windows. Taniguchi et al. (2001) showed that this approach allows freight carriers to reduce their costs of operating pickup/delivery trucks as well as reducing CO_2 emissions. In the future, advanced information systems will be more widely applied in the logistics industry, allowing both efficient and environmentally friendly delivery systems to be realized.

1.2.3 City Planning and City Logistics

Freight transport should be explicitly considered in urban planning. However, many cities have not yet established a master plan for urban freight transport. This is attributed to the lack of appropriate personnel, knowledge, and data for urban freight transport (OECD 2003). Not surprisingly, many cities have no administrative unit that is fully devoted to addressing urban freight issues. In addition, city planning normally considers long-term issues over a period of several years, while the logistics planning horizons of private companies are more short term—often over a few months. This mismatch of the length of planning periods often makes it difficult for freight issues to be addressed effectively by both city authorities and private companies.

Since urban freight transport is very important for sustainable development in urban areas, cities should pay more attention to it including it in their city planning. Each city needs to establish comprehensive visions for urban freight issues. To do this, city planners should gain an appropriate level of knowledge and know-how on urban freight issues and collect necessary data.

1.2.4 Land Use Planning and City Logistics

A common problem in city planning involves the inadequate location of logistics facilities. Freight carriers have difficulty in finding appropriate land for logistics terminals in urban areas because of zoning regulations and land prices. If they build their logistics terminals far from an expressway interchange and if the area between the interchange and the logistics terminals is then developed as a residential area, freight transport generates environmental and safety problems due to large trucks traveling on the roads through the residential area (Taniguchi and Nemoto, 2002). This type of problem can be solved by incorporating freight transport plans within land use planning processes.

1.2.5 Units of Urban Freight Transport Planning

Another problem related to urban freight transport planning is associated with the unit of planning. Basically, each individual authority makes its own urban traffic plans, which are typically oriented toward passenger cars. But the size of cities are often too small for freight transport planning, since freight transport in any one city is only a part of the supply chain that includes line haul transport between cities or even between countries. Therefore, urban freight transport planning in a city should be harmonized with other adjacent cities. Several cities in a wide area, or in a corridor, should preferably discuss their ideas relating to urban freight transport planning to make a common plan for freight transport. Such an institutional framework for urban freight transport planning within a wider area is essential for sustainable development of urban areas.

1.2.6 Subsidies and Additional Charges from the Public

Urban freight transport is a private sector activity undertaken by companies. In principle, neither subsidies nor additional charges should be required to be made by the public sector to private companies. However, freight transport has a large influence on the economic development of cities as well as social and environment systems. In some cases the central government and city authorities assist freight carriers and shippers financially by providing subsidies or low-interest loans. In other cases they impose additional charges to improve the environment. This public commitment is based on the following reasons:

a. It is difficult to internalize the external benefits by new investment (e.g., introducing low-emission trucks by freight carriers).
b. It is hard to internalize the external diseconomy by operating freight vehicles (e.g., congestion charging).
c. A huge initial investment may be required to start a large freight transport project (e.g., building a large-scale logistics terminal for cooperative delivery systems by small and medium size enterprises).

Subsidies are sometimes provided to freight carriers who use low-emission trucks (e.g., compressed natural gas vehicles) in urban areas. Charging vehicles coming into central areas has been considered in many large cities and was implemented in London in 2003. Subsidies are commonly provided for encouraging small and medium size enterprises to participate in innovative logistics projects, such as advanced information systems and cooperative delivery centers.

1.3 INCORPORATING RISKS IN URBAN FREIGHT TRANSPORT

There have been increasing concerns about risks caused by natural hazards including earthquakes, flooding, tsunamis, snowfall, and bushfires, as well as man-made hazards of crashes and terrorist attacks. Although in principle these risks should be well assessed and incorporated in city logistics, they are not fully taken into account in modeling city logistics (Taniguchi et al. 2001) and implementing city logistics schemes in urban areas. The reasons are because (a) assessing the risks related to city logistics is hard due to uncertainty of these events, (b) incorporating risks of natural and man-made hazards incurs additional costs on logistical operations, and (c) natural and man-made disasters are not regarded as being within the logistics manager's responsibility.

Recently, we have encountered extremely destructive disasters generated by the 2011 Tohoku earthquake and tsunami in Japan in 2011; the tsunami after the northern Sumatra earthquake in the Indian Sea region in 2004; Hurricane Katrina in the Gulf of Mexico area, United States, in 2005; the Sichuan earthquake in China in 2008; and bushfires in Melbourne, Australia, in 2009. In addition, there were the chemical attacks using sarin on the subways in Tokyo in 1995; the September 11,

2001, terrorist events in New York and Arlington, Virginia, in the United States; the blast attack in London in 2005; and the piracy attacks off the Somalian coast in 2009 as well as similar ongoing incidents. These threats triggered a change in the mind-sets of stakeholders to start taking into account risks due to natural and man-made hazards.

In private firms these risks are discussed from the viewpoint of business continuity management. We can find a typical case of critical disruption of the supply chain caused by the Niigataken Chuetsu-oki earthquake in 2007, which resulted in cracked walls and ceiling collapse in the main manufacturing plant of a small firm named Renesas that produced micro computers for automobile manufacturers in Japan. The disruption of the Renesas factory generated the shutdown of many automobile assembly factories throughout the world for a week, since small inventory of such parts in just-in-time production systems was so vulnerable to the threats of earthquake and depended too much on supply as a critical part from a single firm. After this, most logistics companies realized the importance of incorporating risks of natural disasters in logistics systems from the point of business continuity management.

In the public sector, risks due to natural and man-made hazards in urban freight transport systems are directly related to public welfare and public health in emergency cases. The public sector is interested in knowing how to mitigate the damage to urban logistics facilities and recover quickly after disasters from the viewpoint of delivering goods needed for maintaining a high quality of life in urban areas.

In this context we need to consider multiple stakeholders involved in urban freight transport systems—namely, shippers, freight carriers, administrators, and residents (consumers)—to incorporate any risks in city logistics. As there are different motivations, objectives, and behaviors among these stakeholders in facing risks, special methodologies, including multiagent models, are required for modeling the behavior of stakeholders.

In general, city logistics takes into account the day-to-day risks of delay when arriving at customers for delivering or picking up goods due to recurrent congestion generated by the concentration of traffic in peak periods, crashes, and sporting events. However, we need to incorporate less frequent but severe effects caused by cyclones, earthquakes, tsunamis, and floods and others extreme weather events. Figure 1.2 illustrates the classification of risks related to city logistics. The horizontal axis shows the source of difficulty for assessing risks caused by events and the vertical axis indicates the frequency of events.

The first level of difficulty comes from complexity. The effects of congestion on the road networks are complex, and it is hard to anticipate travel times due to congestion, since the demand of passenger and

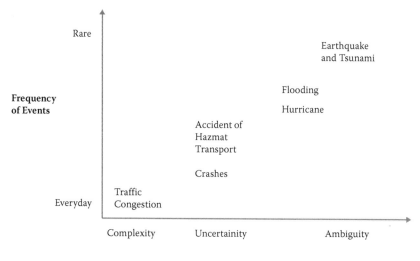

FIGURE 1.2 Classification of risks related to city logistics.

freight traffic fluctuates. If we use vehicle routing and scheduling with time windows (VRPTW) models, conventional VRPTW models can be applied for duplicating the movements of pickup and delivery trucks in urban areas.

The second level of difficulty comes from uncertainty. Crashes on roadways are not predictable and, in particular, vehicles carrying hazardous materials often generate huge damage to residents and buildings as well as other traffic. In this class of difficulty, we need to develop models that take into account the dynamic and stochastic nature of travel times and connectivity of road networks. Then VRPTW-D (-dynamic) (e.g., Taniguchi and Shimamoto 2004) or VRPTW-P (-probabilistic) (e.g., Ando and Taniguchi 2006) models are available to duplicate the dynamic adaptation of starting time and route choice of pickup and delivery trucks in urban areas based on ITS applications.

The third class of difficulty comes from ambiguity. Bad weather conditions and natural hazards are included in this class and the events occur less often but greatly affect urban freight transport systems. In this class, more sophisticated treatments are required for duplicating the behavior of stakeholders. As simply applying VRPTW-D or VRPTW-P models is not sufficient in ambiguous situations, multiobjective models or multiagent models and simulation are effective to assess the effects of these events and evaluate initiatives for responding to them. The ambiguity of the situation is generated by the unpredictable interaction among stakeholders as well as the action–reaction relationships between stakeholders and the environment.

For coping with the risks of complexity, uncertainty, and ambiguity, the concept of risk governance has been proposed (Kröger 2008). The idea of risk governance is beyond risk management and includes the cycle of preassessment, appraisal, characterization and evaluation, and management. The comprehensive framework of risk governance allows us to understand the dependency of each stakeholder related to city logistics initiatives and the critical infrastructures as well as the critical point of supply chains.

Risk has been defined as "the chance of something happening that will have an impact on objectives" (AS/NZS 2004). City logistics aims to reduce the total costs, including economic, social, and environmental, associated with urban goods movement. There are a number of aims and objectives of urban freight systems that are under threat, such as the health and safety of citizens and the drivers of vehicles, the fulfillment of delivery contracts (e.g., city curfews and time windows), and reducing climate change.

There is a need to incorporate uncertainty into models for city logistics to ensure that schemes will perform well into the future. A variety of methods have been used to incorporate uncertainty in supply chain modeling, such as scenario and contingency planning, decision trees, and stochastic programming (Shapiro 2007).

1.4 METHODOLOGY

1.4.1 Robustness

Optimization models that are used to plan, design, and evaluate city logistics schemes often have a number of input variables that are uncertain such as parameters, resources, and operational limits. Mathematical programming can be used to represent systems by an objective function, decision variables, and constraints. Feasible solutions are those that satisfy the constraints of a system. An optimal solution is defined as a feasible solution where the values of the decision variables provide the best value of the objective function.

Solution robustness investigates whether the optimal solution is maintained when there are changes to values in the input data. In particular, solution robustness considers how close the optimal solution is when the input data is changed.

Model robustness considers the effect on feasibility for changes to the input data. This involves determining how close to feasible the optimal solution is when there are changes to the input data.

Robustness of logistics and supply chain networks have received considerable attention recently (Bok and Park 1998; Christopher and Peck 2004; Mo and Harrison 2005; Yu and Li 2000).

Procedures need to be developed for identifying solutions that remain close to optimal and close to feasible when there are changes to the values of the input variables due to their uncertainty. Robust optimization investigates the trade-offs between solution robustness and model robustness (Mulvey, Vanderbei, and Zenios 1995). Several methods have been developed for representing uncertainty in optimization models, including probability distributions, fuzzy logic, and scenario based.

1.4.2 Stochastic Programming

Stochastic programming represents a system as a probabilistic (stochastic) model that explicitly incorporates the coefficients of parameters as random variables within the problem formulation (Birge and Louveaux 1997). This is in contrast to methods such as linear programming, where the coefficients are assumed to be constant, specified by one value. Stochastic programming produces solutions that perform better when parameter coefficients vary from their mean or estimated values.

The possible gain from solving the probabilistic (stochastic) model is measured by the Value of the Stochastic Solution (VSS). This represents the value of knowing and using the distributions of future outcomes. VSS is relevant to problems where the values of parameter coefficients are uncertain and no further information about their future is available. It quantifies the cost of ignoring uncertainty when making a decision based on the model's solution.

In static vehicle routing and scheduling problems, the actual travel time between customers is uncertain; however, a single value estimate (or forecast) is usually made (Psaraftis 1995). Stochastic (probabilistic) models allow random inputs based on probability distributions to be incorporated. Stochastic programming has been applied to the design of supply chain networks (Santoso et al. 2005; Shapiro 2007; Shapiro 2008; Snyder 2006).

To account for the uncertainty of predicting the arrival time of trucks visiting customers, the probabilistic (stochastic) vehicle routing problem with soft time windows (VRPSTW) model involves the estimation of an expected penalty cost (Laporte, Louveaux, and Mercure 1992).

Stochastic programming was used to estimate the benefits (cost savings) of using stochastic programming for vehicle routing and scheduling with time window and variable travel times in urban areas (Thompson et al. 2011).

Late delivery of goods to receivers can lead to missed sales in retailing from stock not being available in shops. Business to Consumer (B2C) distribution can also be disrupted due to delivery failure in home deliveries from persons not being at home, which leads to increased logistics costs. Vehicles running late may not be allowed to enter inner city areas where strict curfews for delivery vehicles have been specified.

Stochastic programming has recently been applied to the design of supply chain networks (Santoso et al. 2005; Shapiro 2007, 2008; Snyder 2006).

1.4.3 Simulation

Simulation modeling provides a useful tool for designing and evaluating urban freight and logistics systems. It has been used widely in the design of loading and unloading facilities and the layout within warehouses and distribution centers. Operational performance measures can be estimated for varying physical designs and demand patterns. This can facilitate the development of contingency plans to cater to extreme conditions.

Micro-simulation was applied to determine a congestion management strategy for the Kallang-Paya Lebar Expressway (KPE) in Singapore (Keenan et al. 2009). The KPE involves a 9 km road tunnel and the level of service was estimated for various traffic levels, including freight vehicles, to ensure reasonable safety levels using VISSIM (Verkehr in Städten—Simulations).

Simulation has also been used to design security and scheduling procedures for reducing the threat of terrorism. A delivery vehicle scheduling system was developed for the planning and operation of the Sydney 2000 Olympics using the Planimate simulation software (Pearson and Gray 2001). Animation of freight activities assisted in facilitating effective communication between key stakeholders, including security forces. This system was used to determine specific time slots of deliveries to Olympic venues and produce a master delivery schedule (MDS). It provided the capability to verify the robustness of the final schedule and to investigate scenarios including terrorist events.

1.4.4 Multiobjective Optimization

There are often several criteria considered simultaneously in supply chain and logistics management. Often, multiple objectives are weighted into a single one to address this issue. However, it is sometimes difficult to set the weights for each objective in advance. Thus, it can typically be modeled and solved within the framework of the multicriteria decision-making problem and the multiobjective optimization problem (MOP) (i.e., multiobjective programming problem). It is common for these problems to not present a unique, optimal solution. Where the values of objective functions are mutually conflicting, all objective functions cannot be simultaneously optimized. In this case, a set of so-called Pareto optimal solutions (e.g., Chancong and Haimes 1983; Sawaragi, Nakayama, and Tanino 1985), which also imply noninferior or nondominated solutions, are determined.

MOP is generally formulated as follows:

$$\min \mathbf{f}(\mathbf{x}) = [f_1(\mathbf{x}), f_2(\mathbf{x}), \ldots, f_m(\mathbf{x})]^T \quad (1.1)$$

$$subject\ to\ x \in S \quad (1.2)$$

where f is a vector of m objective functions to be minimized, **x** is a vector of decision variables, and **S** is a set of all feasible solutions. A solution to the preceding problem, $\mathbf{x}^*$, is called the Pareto optimal solution if and only if $f_i(\mathbf{x}^*) \leq f_i(\mathbf{x}')\ \forall i$ and $f_i(\mathbf{x}^*) < f_i(\mathbf{x}')\ \exists i$. Here, the Pareto optimal solution can be defined as a solution if no objective function value of $\mathbf{x}^*$ is greater than that of $x¢$, and there is at least one objective function value of $\mathbf{x}^*$ that is less than that of $x¢$.

There are, in general, many such optimal solutions, and only a set of Pareto optimal solutions are obtained. Decision makers have to choose a final preferred one from among the Pareto optimal solutions based on personal preferences or additional criteria. There have been a variety of approaches to obtain the Pareto optimal solutions as well as to select the final preferred one, many of which are reviewed in the surveys by Ehrgott and Gandibleux (2002) and Figueira, Greco, and Ehrgott (2005), providing a comprehensive overview of multicriteria decision making and multiobjective optimization.

To facilitate the decision making for supply chain and logistics optimization under the conditions of risk due to the consideration of different demand risks from many different customers, several MOP models have been proposed (e.g., Weber and Ellram 1992; Weber and Current 1993; Ghodsypour and O'Brien 2001; Wu and Olson 2008). Recently, MOP has increasingly been used for a decision-making tool in the field of green supply chain management and reverse logistics, including hazardous material transport and hazardous materials (hazmat) waste management (e.g., Iakovou 2001; Zografos and Androutsopoulos 2004; Chang, Nozick, and Turnquist 2005; List et al. 2006; Yong et al. 2007; Sheu 2008), since these have uncertainty, complexity, and potential risks for people in nature and in the process of transport, transaction, and purchasing.

Both costs and risks in supply chain and logistics management are required to be evaluated simultaneously, and these are often conflicting. When this is the case, total operational costs are commonly estimated, consisting of transport cost, inventory cost, reprocessing cost, and final disposal cost. The assessments are undertaken at the same time for the operational risks generated for transportation, storage, reprocessing, and final disposal.

1.4.5 Multiagent Simulation

Each of the four major stakeholders involved in city logistics—shippers, freight carriers, residents, and administrators—have different objectives and different types of behavior. Shippers attempt to minimize their costs in supply chains. Freight carriers try to meet shippers' requests to collect and deliver goods within strict time windows. Residents want a quiet, noiseless atmosphere and clean air in their community. Administrators aim to stimulate the vitality of a city with sustainable transport systems. Understanding the behavior of the stakeholders and interaction among them is needed for evaluating city logistics measures before implementing them.

Multiagent models generally replicate the behavior and interaction among multiple agents investigated (e.g., Weiss 1999; Ferber 1999; Wooldridge 2002). Urban freight transportation systems can be represented using multiagent modeling techniques, since they are most suitable to understand and study the behavior of stakeholders in urban freight transport systems and their response to policy measures. Davidsson et al. (2005) provided a survey of existing research on agent-based approaches in freight transport and noted that agent-based approaches seem very suitable for this domain. Van Duin, Bots, and van Twist (1998) presented dynamic actor network analysis for complex logistics problems. Ossowski et al. (2005) applied multiagent approaches to decision support systems in traffic management. Jiao, You, and Kumar (2006) presented an agent-based framework in a global manufacturing supply chain network. Published literature highlights a number of interesting examples of multiagent approaches to transport logistics problems, but most of them do not directly focus on urban freight transport systems.

Multiagent models have been developed to investigate the effects of city logistics schemes in which the interactions between shippers, freight carriers, and administrators are considered (Taniguchi, et al. 2007). In this study, the multiagent models include a reinforcement learning process to take better policy in the next step based on the reward given as a result of the previous action of an agent. Here multiagent models were used for evaluating the behavior and interaction among stakeholders who are involved in urban freight transport systems as well as effects of city logistics measures. Multiagent simulation on a small test road network demonstrated that the VRPTW-D model, which dynamically adjusted vehicle routing planning to the current travel times, generated good performance in terms of increasing profits for freight carriers and decreasing costs for shippers. After applying multiagent models on a large test road network, it was observed that by introducing the VRPTW-D model generated a win–win situation by increasing profits for freight carriers and decreasing the costs for shippers. The results also show that the implementation of road pricing can reduce NO_x emissions but may increase the costs for shippers. To avoid such effects, introducing cooperative freight transport systems helps shippers to reduce their costs.

A multiagent model has been developed in which five stakeholders—freight carriers, shippers, residents, administrators, and urban motorway operators—were considered (Tamagawa et al., 2009). This application embedded the Q-learning process in decision making of policy by agents, taking into account the reward of previous action. After applying multiagent models in an urban road network, it examined the performance of several city logistics measures including road pricing and truck bans. It was found that despite the implementation of the city logistics measures and the increased demand for urban motorway use, freight carriers could maintain their transport costs at an equal level without any city logistics measures or tolling, and they could also keep the delivery charge at the same level. As a result, shippers could keep their delivery cost at the same level as before. It was concluded that the implementation of these measures led to an improved environment for all stakeholders.

Hybrid models based on aggregate macroeconomic interactions, discrete event microsimulation, and agent-based modeling were developed to represent urban goods movement (Donnelly, 2009). This approach was used in the practical domain of road network planning in Portland for examining some city logistics scenarios using existing data sets.

Teo, et al. (2012) presented a multiagent model for evaluating city logistics policy measures of road pricing and load factor control using auctioning between shippers and freight carriers. The result was that a distance-based road pricing scheme, with a load factor control scheme, has the potential to provide a win-win situation that meets the objectives of the key stakeholders in city logistics.

1.5 HEALTH

1.5.1 Introduction

Freight vehicles operating in urban areas produce a substantial amount of noise and emissions that can be harmful to personal health. Carriers need to consider the health of residents as part of their Corporate Social Responsibility (CSR). This can involve operating alternative fuel vehicles that produce less harmful emissions as well as encouraging good driving practices such as ecodriving.

Personal Health is an important element of the vision for city logistics (Taniguchi, Thompson, and Yamada 2004). The health and safety of employees of freight carriers is a substantial issue and there is increasing pressure on employers to be more proactive with respect to protecting the health of their employees. This section outlines approaches used to model the effects of air quality and physical activity of drivers.

1.5.2 Air Quality

Drivers of trucks and vans are often exposed to high levels of emissions for extended periods. A number of research studies have shown that there are higher levels of air pollution inside road vehicles compared to ambient air quality (Chertok 2004; Fruin 2004; Rodes et al. 1998). The level of in-cabin exposure to air pollutants is a function of the time a person spends in the cabin in urban streets. Therefore, drivers have an increased risk of cancer and respiratory diseases such as asthma.

There are a number of measures that are used in public health to quantify the effects of diseases. The disability adjusted life year (DALY) is an indicator of the burden of disease (BoD) in the community. One DALY represents one lost year of "healthy" life and is a combination of years of life lost (YLL) as a result of premature mortality plus an equivalent number of "healthy" years of life lost as a result of disability (YLD). DALYs are the summation of years due to premature mortality (YLL) in the population and the years lost due to disability (YLD). DALYs are regularly determined for particular states and cities and typically derived for particular diseases such as diabetes mellitus and cardiovascular disease (Murray and Lopez 1996).

Improved methods are needed to develop and identify cost-effective methods to reduce the risk of health problems for drivers from air pollution. There is a need to build public health impact evaluation using tools, such as the DALY health measurement, into city logistics planning for avoidable deaths. Kayak and Thompson (2007) report that the possible contribution to the BoD for the population of the Melbourne statistical division (MSD) jurisdictions from exposure to diesel fuel emissions by in-vehicle diesel engine environments is multiples greater than that to the population outside the vehicles.

1.5.3 Physical Activity

Rising levels of obesity are becoming a serious health problem in many countries. Residents in many cities are leading increasingly sedentary lifestyles. Physical activity is most commonly undertaken at work or home or during recreation and transport.

The current Australian physical activity guidelines are to "put together at least 30 minutes of moderate-intensity physical activity on most, preferably all, days" (Commonwealth Department of Health and Family Services 1998). However, drivers face increased risks of not undertaking sufficient levels of physical activity at work due to the extended periods of sitting while working.

There is a need to incorporate more physical activity in urban distribution since many logistics tasks have been mechanized in the last fifty years. A significant amount of physical activity has been taken out of contemporary daily distribution systems in cities, with typical motorization of delivery of newspapers and mail as well as the collection of garbage. Drivers often experience long periods sitting in vehicles while driving or waiting. A classic epidemiological study of bus conductors in London concluded that the higher physical activity of conductors on double-decker buses in London attributed to the lower disease incidence and mortality for the conductors compared with drivers (Morris et al. 1953).

A person's weight is largely determined by the amount of energy expenditure and food energy intake. By combining levels of predicted energy expenditure (including that from nonmotorized distribution) with dietary details, the weight of individuals can be simulated (Westerterp et al. 1995; Payne and Dugdale 1977). The health benefits can then be determined since many disease studies relate the risk of chronic diseases to body mass index (BMI).

Daily physical activity levels (PALs) can be determined by estimating the total energy expenditure, expressed as a multiple of the basal metabolic rate (Ainsworth et al. 2000). It is recommended that average PALs should be above 1.6 (AICR 2007).

Estimating energy expenditure for an individual over a daily period is a complex and challenging task. There are several methods of calculating physical activity levels. A simple method has developed for determining basal, activity, and total energy expenditure levels based on personal attributes such as age, height, weight, and gender as well the duration and metabolic rate of activities undertaken (Ainsworth et al. 2000; Gerrior, Juan, and Basiotis 2006). Accelerometers can be used to measure accurately the amount of energy expenditure undertaken due to physical activity. Activity diaries can also be used to estimate the frequency, intensity, and duration of physical activities undertaken by individuals.

Recently, a growing proportion of mail deliveries within Australian cities have been undertaken using trolleys and bicycles instead of motorcycles and vans. This is partially due to the difficulty in recruiting licensed motorcycle riders as well as the increasing number of motorcycle riders that are becoming overweight. Mail deliveries to dwellings in residential areas by bicycle has many health benefits for riders. In addition, there are environmental benefits of less air pollution as well as social benefits associated with safety costs associated with crashes.

1.6 HUMAN SECURITY ENGINEERING

1.6.1 Introduction

A range of natural and man-made disasters can cause disruptions to urban distribution systems. After the emergency response and relief phases, urban logistics systems often need to be reorganized as reconstruction and recovery efforts are undertaken.

The need to build more resilience into urban transport systems to limit the effect of disasters has been well recognized (Murray-Tuite and Mahmassani 2004; Murray-Tuite 2006). City logistics schemes can provide an efficient means of continuing distribution services when the capacity of the urban traffic system has been reduced as a result of disasters.

The capacity of traffic links is often lost or reduced following a disaster, resulting in increased travel time, delays, and delay penalties. Origin and destination patterns for freight vehicle trips can change due to links being blocked, leading to changes in routes within urban areas.

There is a need to design appropriate city logistics schemes to operate throughout the reconstruction period following a disaster. Models can assist in identifying vulnerable links as well as determining the most efficient schedule of reconstruction projects.

The direct losses, including the reconstruction of traffic links and traffic management, as well as the indirect losses such as business disruption and increased delay, can be estimated using economic loss modeling (Buckle 2005).

Financial loss from disasters can best be predicted using catastrophe models (Grossi and Kunreuther 2005). Hazard and inventory information can be used to estimate the vulnerability of structures to damage from a disaster. Losses are predicted by direct costs of repair or reconstruction as well as indirect costs such as business interruption and evacuation.

Information of hazards involve consideration of the location, frequency, and severity of disasters, largely based on analysis of historical data. Information concerning the type and strength of man-made structures is required for inventory analysis. Vulnerability modeling can be used to predict the damage to a structure for given disaster events. Loss models can be used to provide estimates of the financial loss incurred from the physical damage.

Estimates of postdisaster trip demands are used to identify vulnerable transport links (including bridges within the Risk from Earthquake Damage to Roadway Systems (REDARS) model (Werner et al. 2005, 2007). This model provides decision guidance to reduce risks and can assist in developing response and recovery strategies.

1.6.2 Man-Made Disasters

The topic of man-made disasters such as war, terrorism, epidemics and pandemics, traffic accidents, nuclear accidents, food or water contamination, and the collapse of buildings provides many challenges for city logistics. In contrast to natural disasters, man-made disasters can occur anywhere there is human activity, while natural disasters are generally regional (Tansel, 1995). It has also been recognized that hazardous waste is the most serious environmental problem among man-made disasters confronting risk managers in the 1990s. Risk management in hazardous material transport has therefore been a major research topic relating to man-made disasters (e.g., Gopalan et al. 1990; List et al. 1991; Beroggi and Wallace 1995; Nozick, List, and Turnquist 1997; Miller-Hooks and Mahmassani 1998).

Since September 11th, 2001, there has been an increasing need to secure transportation infrastructure further due to terror threats and incidents. Supply chains are particularly vulnerable to intentional or accidental disruptions and suggest a multisupplier strategy as a possible approach for alleviating the vulnerability (Sheffi, 2001). Quantitative models that deal with supply chain risks and classify supply chain risks into two types: operational risks and disruption risks (Tang, 2006). Operational risks refer to the inherent uncertainties that inevitably exist in supply chains, such as uncertain customer demands and uncertain costs. Robust supply chain strategies for mitigating disruption risks that would enhance efficiency and resiliency have been proposed. A real-time supply chain management model based on multiagent simulation, with its application assuming bird flu (avian influenza) or terrorist attacks, has been developed (Lau et al. 2008). The risks faced by the poultry supply chain in the epidemic of avian influenza have also been investigated including identifying risk factors, losses and gains, and mitigation strategies used by different players in the supply chain (Mohan et al. 2009).

Surface freight transport, are inherently vulnerable to terrorism. Plant (2004) examines the impact of the terrorist attacks on the North American rail industry and government agencies concerned with railroad security. The usage of real-time information systems and categorization of some intelligent transport subsystems, that can help to protect transport infrastructure and systems against transport terrorism, has also been reviewed (Okonweze and Nwagboso, 2004). A framework for evaluating risks associated with the loss of capacity between origin and destination (OD) on the road transport network from direct targeting by terrorists has also been developed (Murray-Tuite, 2007).

1.6.3 Hazardous Material Transport

Transportation of hazardous goods within urban areas is an important issue in road management. If a truck carrying hazardous material is involved in a crash, it could have a significant impact on people and buildings and other traffic from the explosion or spillage. Information and Communication Technology (ICT) and ITS can be used to assess and manage these risks. Advanced modeling techniques are required for assessing risks and evaluating measures to improve the management of hazardous goods transport in urban traffic networks.

Different risk models can produce different "optimal" paths for a hazmat shipments between a given origin–destination (Erkut and Verter, 1998). Five categories of risk models have been identified: (1) traditional risk, (2) population exposure, (3) incident probability, (4) perceived risk, and (5) conditional risk.

Multiobjective optimization models can be used to incorporate risks associated with hazardous goods transport. A multiobjective programming model incorporating total costs, total perceived risk, individual perceived risk, and individual disutility has been developed (Giannikos, 1998). A method for identifying nondominated paths for multiple routing objectives in networks where the routing characteristics are uncertain, and the probability distributions that describe those attributes vary by time of day, was developed by Chang et al. (2005). It was found that a mix of routes consisting of the set of safest routes, and the safest share of traffic between these routes, leads to better risk-averse strategy based on a game theoretic approach (Bell, 2006). Beroggi (1994) presented a real-time routing model for assessing the costs and risks in a real-time environment and highlighted that the ordinal preference model turned out to be superior to the utility approach. A multiobjective optimization model of hazmat delivery systems, incorporating exposure risks of traffic accidents to residents in urban areas was proposed by Pradhananga et al. (2009). Here, the set of Pareto optimal solutions provides a number of economic and safe routing solutions.

1.6.4 Traffic Safety

Research relating to traffic safety and accidents has been largely focused on empirical studies aimed at identifying the relationships between: crash frequency and traffic, road factors including vehicle kilometers, average hourly traffic volume per lane, average occupancy, lane occupation, average speed, its standard deviation, curvature, road geometry, and ramp section design (e.g., Jovanis and Chang 1986; Miaou and Lum 1993; Shankar, Mannering, and Barfield 1995; Abdel-Aty and Radwan 2000). Relationships between accident rates and traffic characteristics, including hourly traffic volume, level of service, weather, and hourly traffic flow of cars and trucks, have been determined (e.g., Fridstrøm et al. 1995; Ivan, Pasupathy, and Ossenbruggen 1999; Martin 2002). Driver behavior has also been identified as a major contributing factor for traffic accidents. Sleep-related or sleepiness-related driving accidents have been investigated by Horne and Reyner (1995) and Sagberg (2001).

The effect of road freight vehicles and heavy goods vehicles (HGVs) on traffic accidents has recevied very little attention (Hiselius 2004; Ramírez et al. 2009). The relationship of ramp section design and truck accident rates was investigated by Zaloshnja and Miller (2004). An interview questionnaire survey of professional drivers, including truck drivers was used to identify factors contributing to the probability of falling asleep and crash risk (Tzamalouka et al. 2005). Applications of intelligent transport systems for the prevention of traffic accidents caused by freight traffic have also been investigated (Palkovics and Fries 2001; Sarvi and Kuwahara 2008).

1.7 CONCLUSION

The three pillars of visions for city logistics, mobility, sustainability, and livability, braced by values such as safety, security, and economic prosperity, should inspire researchers and practitioners to develop and implement solutions for solving urban freight problems.

Contemporary urban logistics systems provide a wide range of benefits for residents. However, significant negative impacts can arise.

There are a number of promising schemes that have the potential to realize the visions of city logistics fully, including:

a. Establishing effective partnerships between key stakeholder groups
b. Implementing information and communication technology and intelligent transport systems
c. Promoting corporate responsibility
d. Incorporating urban freight transport as an integral component of urban planning

There is a need for transport logistics models to incorporate risk so that urban distribution systems can become more resilient with respect to natural and man-made hazards. Models that take risks into account can assist in designing city logistics schemes to improve the health and safety of persons involved in transport logistics as well as residents. The challenge is to discover innovative schemes and develop planning processes that will allow the visions of city logistics to be achieved.

This chapter has described the links between human security engineering and city logistics. The need for models to assist in the recovery of transport infrastructure for the public sector, as well as the need to develop plans for business continuity management for the private sector, was outlined.

Due to growing urbanization and the increasing prevalence of extreme weather events, as well the continuing threat of terrorism, improved urban freight models are required to minimize the disruptions to urban freight systems from natural and man-made hazards.

REFERENCES

Abdel-Aty, M. A. and Radwan, A. E. (2000). Modeling traffic accident occurrence and involvement. *Accident Analysis & Prevention* 32 (5): 633–642.

AICR (2007). *Food, nutrition, physical activity, and the prevention of cancer: A global perspective.* World Cancer Research Fund/American Institute for Cancer Research, Washington, DC: AICR.

Ainsworth, B. E., Haskell, W. L., Whitt, M. C., Irwin M. L., Swartz, A. M., Strath, S. J., O'Brien, W. L., Bassett, D. R. Jr., Schmitz K. H., Emplaincourt, P. O., et al. (2000). Compendium of physical activities: An update of activity codes and MET intensities. *Medicine & Science in Sports & Exercise* 32: S498–S516.

Ando, N., and Taniguchi, E. (2006). Travel time reliability in vehicle routing and scheduling with time windows. *Networks and Spatial Economics* 6 (3–4): 293–311.

AS/NZS (2004). Risk management. AS/NZS 4360, Standards Australia/ Standards New Zealand, Sydney.

Bell, M. G. H. (2006). Mixed route strategies for the risk-averse shipment of hazardous materials. *Networks and Spatial Economics* 6: 253–265.

Beroggi, G. E. G. (1994). A real-time routing model for hazardous materials. European *Journal of Operational Research* 75: 508–520.

Beroggi, G. E. G., and Wallace, W. A. (1995). Operational control of the transportation of hazardous materials: An assessment of alternative decision models. *Management Science* 41 (12): 1962–1977.

Birge, J. R., and Louveaux, F. (1997). *Introduction to stochastic programming*. New York: Springer.

Bok, J. H. L., and Park, S. (1998). Robust investment model for long-range capacity expansion of chemical processing networks under uncertain demand forecast scenarios. *Computers and Chemical Engineering* 22 (7): 1037–1049.

Brandenburger, A. M., and Nalebuff, B. J. (1996). *Co-opetition*. New York: Doubleday.

British Library (1994). *Quality of life in cities—An overview and guide to the literature*. London: World Health Organization and the World Bank.

Buckle, I. (2005). Protecting critical infrastructure systems. CAE Resilient Infrastructure Conference, Rotorua.

Chancong, V., and Haimes, Y. Y. (1983). *Multiobjective decision making; theory and methodology*. Mineola, NY: Dover Publications.

Chang, T. S., Nozick, L. K., and Turnquist, M. A. (2005). Multi-objective path finding in stochastic dynamic networks, with application to routing hazardous materials shipments. *Transportation Science* 39 (3): 383–399.

Chertok, M. (2004). Comparison of air pollution exposure for five commuting modes in Sydney—Car, train, bus, bicycle, and walking. *Health Promotion Journal of Australia* 15 (1): 63–67.

Christopher, M., and Peck, H. (2004). Building the resilient supply chain. *International Journal of Logistics Management* 15 (2): 1–13.

Commonwealth Department of Health and Family Services. (1998). Developing an active Australia: A framework for action for physical activity and health. Canberra.

Cooper, J. (1991). Innovation in logistics: The impact on transport and the environment. In *freight transport and the environment*, ed. M. Kroon, R. Smit, and J. van Ham, 235–254. New York: Elsevier.

Davidsson, P., Henesey, L., Ramstedt, L., Tornquist, J., and Wernstedt, F. (2005). An analysis of agent-based approaches to transport logistics. *Transportation Research Part C* 13C: 255–271.

Donnelly, R. (2009). A hybrid micro-simulation model of freight transport demand. PhD dissertation, the University of Melbourne.

Ehrgott, M., and Gandibleux, X. (2002). *Multiple criteria optimization: State of the art annotated bibliographic surveys*. International Series in Operation Research and Management Science, 52. Boston: Kluwer Academic Publishers.

Erkut, E., and Verter, V. (1998). Modeling of transport risk for hazardous materials. *Operations Research* 46 (5): 625–642.

Ferber, J. (1999). *Multi-agent systems—An introduction to distributed artificial intelligence*. London: Addison–Wesley.

Figueira, J., Greco, S., and Ehrgott, M. (2005). *Multiple criteria decision analysis, state of the art surveys*. International Series in Operation Research and Management Science, 78. New York: Springer.

Fridstrøm, L., Ifver, J., Ingebrigtsen, S., Kulmala, R., and Thomsen, L. (1995). Measuring the contribution of randomness, exposure, weather, and daylight to the variation in road accident counts. *Accident Analysis & Prevension* 27 (1): 1–20.

Fruin, S. (2004). The importance of in-vehicle exposures. ftp://ftp.arb.ca.gov/carbis/research/health/healthup/dec04-1.pdf (retrieved December 5, 2007).

Gerrior, S., Juan, W., and Basiotis, P. (2006). An easy approach to calculating estimated energy requirements, *Prevention of chronic disease*, vol. 3, no. 4. http://www.pubmedcentral.nih.gov/articlerender.fcgi?artid = 1784117 (retrieved May 30, 2008).

Ghodsypour, S. H., and O'Brien, C. (2001). The total cost of logistic in supplier selection, under conditions of multiple sourcing, multiple criteria and capacity constraint. *International Journal of Production Economics* 73: 15–27.

Giannikos, I. (1998). A multi-objective programming model for locating treatment sites and routing hazardous wastes. *European Journal of Operational Research* 104: 333–342.

Gopalan, R., Kolluri, K. S., Batta, R. and Karwan, M. H. (1991). Modeling equity of risk in the transportation of hazardous materials. *Operations Research* 38 (6): 961–973.

Grossi, P., and Kunreuther, H. (2005). *Catastrophe modeling: A new approach for managing risk*. New York Springer–Verlag.

Hiselius, L. W. (2004). Estimating the relationship between accident frequency and homogeneous and inhomogeneous traffic flows. *Accident Analysis & Prevention* 36: 985–992.

Horne, J., and Reyner, L. (1995). Sleep related vehicle accidents. *British Medical Journal* 310: 565–567.

Iakovou, E. T. (2001). An interactive multiobjective model for the strategic maritime transportation of petroleum products: Risk analysis and routing. *Safety Science* 39: 19–29.

Ivan, J. N., Pasupathy, R. K., and Ossenbruggen, P. J. (1999). Differences in causality factors for single and multi-vehicle crashes on two-lane roads. *Accident Analysis & Prevention* 31 (6): 695–704.

Jiao, J., You, X., and Kumar, A. (2006). An agent-based framework for collaborative negotiation in the global manufacturing supply chain network. *Robotics and Computer-integrated Manufacturing* 22: 239–255.

Jovanis, P., and Chang, H. L. (1986). Modeling the relationship of accidents to miles traveled. *Transportation Research Board* 1068: 42–51.

Kayak, H., and Thompson, R. G. (2007). Estimating the health impact of diesel fuel use from road based transport. 29th Conference of Australian Institutes of Transport Research (CAITR), December 5–7, 2007, Transport Systems Centre, University of South Australia, Adelaide.

Keenan, D., Leong, Y. F., Chong, I., Pelosi, S., and Cooper, T. (2009). Singapore Kallng-Paya Lebar Expressway (KPE). *Traffic Engineering and Control* April 2009, 1–5. www.tecmagazine.com

Kröger, W. (2008). Critical infrastructures at risk: A need for a new conceptual approach and extended analytical tools. *Reliability Engineering and System Safety* 93 (12): 1781–1787.

Laporte, G., Louveaux, F. V., and Mercure, H. (1992). The vehicle routing problem with stochastic travel times. *Transportation Science* 26: 161–170.

Lau, H. C., Agussurja, L., and Thangarajoo, R. (2008). Real-time supply chain control via multi-agent adjustable autonomy. *Computers & Operations Research* 35 (11): 3452–3464.

List, G. F., Mirchandani, P. B., Turnquist, M. A., and Zografos, K. G. (1991). Modeling and analysis for hazardous materials transportation: Risk analysis, routing/scheduling and facility location. *Transportation Science* 25 (2): 100–114.

List, G. F., Wood, B., Turnquist, M. A., Nozick, L. K., Jones, D. A., and Lawton, C. R. (2006). Logistics planning under uncertainty for disposition of radioactive wastes. *Computers & Operations Research* 33: 701–723.

Martin, J. L. (2002). Relationship between rate and hourly traffic flow on interurban motorways. *Accident Analysis & Prevention* 34: 619–629.

Miaou, S. P., and Lum, H. (1993). Modeling vehicle accident and highway geometric design relationships. *Accident Analysis & Prevention* 25 (6): 689–709.

Miller-Hooks, E., and Mahmassani, H. S. (1998). Optimal routing of hazardous materials in stochastic, time-varying transportation networks. *Transportation Research Record* 1645: 143–151.

Mo, Y., and Harrison, T. P. (2005). A conceptual framework for robust supply chain design under uncertainty. In *Supply chain optimization*, ed. J. Geunes and P. M. Pardalos, chap. 8. Dordrecht, The Netherlands: Kluwer Academic Publishers.

Mohan, U., Viswanadham, N., and Trikha, P. (2009). Impact of avian influenza in the Indian poultry industry: A supply chain risk perspective. *International Journal of Logistics System and Management* 5: 89–105.

Morris J. N., Heady, J. A., Raffle, P. A. B., Roberts, C. G., and Parks, J. W. (1953). Coronary heart disease and physical activity of work. *Lancet* 262: 1053–1108.

Mulvey, J. M., Vanderbei, R. J., and Zenios, S. A. (1995). Robust optimization of large-scale systems. *Operations Research* 43: 264–281.

Murray, C. J. L., and Lopez, A. D., eds. (1996). *Summary: The global burden of disease: A comprehensive assessment of mortality and disability from disease, injuries, and risk factors in 1990 and projected to 2020*. Cambridge: Harvard University (on behalf of the World Health Organization and the World Bank).

Murray-Tuite, P. M. (2006). A comparison of transportation network resilience under simulated system optimum and user equilibrium conditions. Proceedings 2006 Winter Simulation Conference, IEEE 1-4244-0501-7/06.

———. (2007). A framework for evaluating risk to the transportation network from terrorism and security policies. *International Journal of Critical Infrastructure* 3: 389–407.

Murray-Tuite, P. M., and Mahmassani, H. S. (2004). Methodology for the determination of vulnerable links in a transportation network. *Transportation Research Record* 1882: 88–96.

Nozick, L. K., List, G. F., and Turnquist, M. A. (1997). Integrated routing and scheduling in hazardous materials transportation. *Transportation Science* 31 (3): 200–215.

OECD Working Group on Urban Freight Logistics. (2003). Delivering the goods: 21st century challenges to urban goods transport. OECD.

Okonweze, A., and Nwagboso, C. (2004). Securing transport systems and infrastructure against terrorism using ITS. *Proceedings of International Conference of Information and Communication Technologies: From Theory to Applications*, 59.

Ossowski, S., Hernandez, J. Z., Belmonte, M-V., Fernandez, A., Garcia-Serrano, A., Perez-de-la-Cruz, J-L., Serrano, J-M., and Triguero, F. (2005). Decision support for traffic management based on organizational and communicative multiagent abstractions. *Transportation Research Part C* 13C: 272–298.

Palkovics, L., and Fries, A. (2001). Intelligent electronic systems in commercial vehicles for enhanced traffic safety. *Vehicle System Dynamics* 35 (4–5): 227–289.

Payne, P. R., and Dugdale, A. E. (1977). A model for the prediction of energy balance and body weight. *Annals of Human Biology* 4(6): 525–535.

Pearson, L., and Gray, R. (2001). Simulation software empowers logistics success at Sydney 2000 Olympics. SimTec Conference 2001, Simulation Industry Association Australia (SIAA).

Plant, J. F. (2004). Terrorism and the railroads: Redefining security in the wake of 9/11. *Review of Policy Research*, 21 (3): 293–306

Psaraftis, H. N. (1995). Dynamic vehicle routing: status and prospects. *Annals of Operations Research* 61: 143–164.

Ramírez, B. A., Izquierdoa, F. A., Fernández, C. G., and Méndez, A. G. (2009). The influence of heavy goods vehicle traffic on accidents on different types of Spanish interurban roads. *Accident Analysis & Prevention* 41: 15–24.

Rodes, C., Sheldon, L., Whitaker, D., Clayton, A., Fitzgerald, K., Flanagan, J., et al. (1998). *Measuring concentrations of selected air pollutants inside California vehicles*. Sacramento, CA: California Air Resources Board.

Sagberg, F. (2001). Road accident caused by drivers falling asleep. *Accident Analysis & Prevention* 33: 31–41.

Santoso, T., Ahmed, S., Goetschalckx, M., and Shapiro, A. (2005). A stochastic programming approach for supply chain network design under uncertainty. *European Journal of Operational Research* 167: 96–115.

Sarvi, M., and Kuwahara, M. (2008). Using ITS to improve the capacity of freeway merging sections by transferring freight vehicles. *IEEE Transactions—Intelligent Transportation Systems* 9 (4): 580–588.

Sawaragi, Y., Nakayama, H., and Tanino, T. (1985). *Theory of multiobjective optimization*. Orlando, FL: Academic Press.

Shankar, V., Mannering, F., and Barfield, W. (1995). Effect of roadway geometrics and environmental factors on rural freeway accidents frequencies. *Accident Analysis &. Prevention* 27 (3): 371–389.

Shapiro, A. (2007). Stochastic programming approach to optimization under uncertainty. *Mathematical Programming Series A and B* 112 (1): 183–220.

Shapiro, J. F. (2008). *Modeling the supply chain,* 2nd ed. Belmont, CA: Duxbury, Thomson.

Sheffi, Y. (2001). Supply chain management under the threat of international terrorism. *International Journal of Logistics Management* 12 (2): 1–11.

Sheu, J. B. (2008). Green supply chain management, reverse logistics and nuclear power generation. *Transportation Research Part E* 44: 19–36.

Snyder, L. V. (2006). Facility location under uncertainty: A review. *IIE Transactions* (38) 7: 537–554.

Tamagawa, D., Taniguchi, E., and Yamada, T. (2009). Evaluating city logistics measures using the multi-agent model. *Proceedings of the 6th International Conference on City Logistics,* Puerto Vallarta.

Tang, C. S. (2006). Perspectives in supply chain risk management. *International Journal of Production Economics* 103: 451–488.

Taniguchi, E., and Nemoto, T. (2002). Transport demand management for freight transport. In *Innovations in freight transport,* ed. E. Taniguchi and R. G. Thompson, 101–124. Southampton, England: WIT Press.

Taniguchi, E., and Shimamoto, H. (2004). Intelligent transportation system based dynamic vehicle routing and scheduling with variable travel times. *Transportation Research Part C* 12C (3–4): 235–250.

Taniguchi, E., and Thompson, R. G. (2002a). Modeling city logistics. *Transportation Research Record, Journal of the Transportation Research Board* 1790: 45–51.

———, eds. (2002b). *Innovations in freight transport.* Southampton, England: WIT Press.

Taniguchi, E., Thompson, R. G., and Yamada, T. (1998). Vehicle routing and scheduling using ITS. *Proceedings of 5th World Congress on Intelligent Transport Systems,* VERTIS, Seoul.

——— (2001). Recent advances in modeling city logistics. In *City logistics II,* ed. E. Taniguchi and R. G. Thompson, 3–34. Kyoto: Institute of Systems Science Research.

——— (2004). Visions for city logistics. In *Logistics systems for sustainable cities,* ed. E. Taniguchi and R. G. Thompson, 1–16. New York: Elsevier.

Taniguchi, E., Thompson, R. G., Yamada, T., and van Duin, R. (2001). *City logistics: Network modeling and intelligent transport systems.* Oxford, England: Pergamon.

Taniguchi, E., Yamada, T., and Okamoto, M. (2007). Multiagent modeling for evaluating dynamic vehicle routing and scheduling systems. *Journal of the Eastern Asia Society of Transportation Studies* 7: 933–948.

Tansel, B. (1995). Natural and man-made disasters: Accepting and managing risks. *Safety Science* 20: 91–999.

Teo, J.S.E., Taniguchi, E. and Qureshi, A.G. (2012) Evaluation of distance-based and cordon-based urban freight road pricing on e-commerce environment with multi-agent model, *Transportation Research Record: Journal of the Transportation Research Board,* Washington, D.C. 2269, 127–134.

Thompson, R. G., and Taniguchi, E. (2001). City logistics and freight transport. In *Handbook of logistics and supply chain management,* ed. A. M. Brewer, K. Button, and D. A. Hensher, 393–404, Handbooks in Transport, vol. 2. Oxford: Elsevier.

Thompson, R.G., E. Taniguchi and T. Yamada (2011). Estimating benefits of considering travel time variability in urban distribution, *Transportation Research Record: Journal of the Transportation Research Board*, Washington, D.C., No. 2238, 86–96.

Tzamalouka, G., Papadakaki, M., and El Chliaoutakis, J. (2005). Freight transport and non-driving work duties as predictors of falling asleep at the wheel in urban areas of Crete. *Journal of Safety Research* 36: 75–84.

van Duin, J. H. R., Bots, P. W. G., and van Twist, M. J. W. (1998). Decision support for multi-stakeholder logistics. *Proceedings TRAIL Conference*, Scheveningen.

Weber, C. A., and Current, J. R. (1993). A multiobjective approach to vendor selection. *European Journal of Operational Research* 68: 173–184.

Weber, C. A., and Ellram, L. M. (1992). Supplier selection using multi-objective programming: A decision support system approach. *International Journal of Physical Distribution and Logistics Management*, 23 (2): 3–14.

Weiss, G., ed. (1999). *Multiagent systems—A modern approach to distributed artificial intelligence.* Cambridge, MA: MIT Press.

Werner, S. D., Cho, S., Taylor, C. E., and Lavoie, J.-P. (2007). Use of seismic risk analysis of roadway systems to facilitate performance-based engineering and risk-reduction decision making. US–Japan Bridge Conference.

Werner, S. D, Taylor, C. E., Cho, S., Lavoie, J-P., Huyck, C. K., Eitzel, C., Eguchi, R. T., and Moore, J. E. (2005). New developments in seismic risk analysis of highway systems. *Seismic Vulnerability of the Highway System* 221–238.

Westerterp, K. R., Donkers, J. H. H. L. M., Fredrix, E. W. H. M., and Boekhoudt, P. (1995). Energy intake, physical activity and body weight: A simulation model. *British Journal of Nutrition* 73: 331–341.

Wooldridge, M. (2002). *An introduction to multiagent systems.* New York: John Wiley & Sons.

Wu, D., and Olson, D. L. (2008). Supply chain risk, simulation, and vendor selection. *International Journal of Production Economics* 114: 646–655.

Yong, Z., Xuhong, L., Haijun, M., and Chenglin, M. (2007). Designing a hazardous waste reverse logistics network for third party logistics supplier under the fuzzy environment. *IEEE International Conference on Service Operations and Logistics Information,* 1–6.

Yu, C-S., and Li, H-L. (2000). A robust optimization model for stochastic logistics problems. *International Journal of Production Economics* 64: 385–397.

Zaloshnja, E., and Miller, T. R. (2004). Costs of large truck-involved crashes in the United States. *Accident Analysis & Prevention* 36: 801–808.

Zografos, K. G., and Androutsopoulos, K. N. (2004). A heuristic algorithm for solving hazardous materials distribution problems. *European Journal of Operational Research* 152 (2): 507–519.

CHAPTER 2

Transport and Logistics in Asian Cities

Tien Fang Fwa

CONTENTS

2.1	Introduction	31
2.2	Urbanization, Freight Transportation, and City Logistics	32
	2.2.1 Consequences of Urbanization in Asia	32
	2.2.2 Characteristics of Urban Freight Transport and Logistics	35
2.3	Impacts of Transport and Logistics on City Development in Asia	38
	2.3.1 Travel Safety	38
	2.3.2 Social Impacts	39
	2.3.3 Environmental Impacts	40
2.4	Need for Incorporating City Logistics into Transport and Land Use Planning	41
2.5	Emergence of the Megalopolis	44
2.6	Opportunities and Challenges of City Logistics Developments in Asian Cities	47
	2.6.1 Enhancing Transport and Land Use Planning Incorporating City Logistics Needs	47
	2.6.2 Exploiting Alternative Transport and Distribution Modes for Freight	48
	2.6.3 Achieving Environmentally Friendly City Logistics for Sustainable City Development	49
References		50

2.1 INTRODUCTION

In the history of human civilization, for many thousands of years before the era of industrialization and the advent of motorization, there was little difference in size and speed between the carriers for people and those for goods. Because of the relatively slow pace of economic activities and low speed of transport movements, traffic congestion, if any, was not of major consequence and the conflict between passenger and goods transport was almost nonexistent.

With the rapid growth in industrialization and motorization along with intensification of economic activities in the second half of the twentieth century, cities tended to grow in size spatially and in population in the continuing process of urbanization. Today, a quarter of all world exports come from East Asia (Stone 2008). The increased urban population and economic activities have led to a steady increase in travel demand within the city, resulting in traffic congestion as traffic growth outpaces the upgrading of transportation infrastructure and implementation of effective traffic management measures. Traffic volume soon surpasses the capacity of the road network, which was not originally designed to cope with the high speed of traffic growth. This is the common problem encountered by major cities in the world, including practically all cities in the developed and developing countries of Asia.

Traffic congestion in the city has created mobility and access problems for passenger and freight transport alike. This chapter addresses the issues and challenges of urban freight transport and city logistics in Asian cities by examining the common problems encountered by the sector in selected cities.

2.2 URBANIZATION, FREIGHT TRANSPORTATION, AND CITY LOGISTICS

2.2.1 Consequences of Urbanization in Asia

The rapid pace of industrialization and motorization in the twentieth century, the last forty years of the century in particular, has facilitated globalization through international trading, and interaction has intensified the process of urbanization. Urbanization refers to the growth process of an urban area into adjoining areas due to either an increase of population or physical expansion of urban infrastructure to accommodate increased economic or social activities, or both. Due to the high concentration of population and economic activities, cities today as an urban system have become increasingly complex by serving as regional hubs for economic development, as well as centers of finance, culture, policies, and politics.

The global proportion of urban population rose dramatically in the second half of the twentieth century as a result of increased industrialization and international trading, which can be largely attributed to the significant technological advances in the freight transportation sector. The United Nations (2005) reported that the proportion of global urban population rose from 29% (732 million) in 1950 to 49% (3.2 billion) in 2005. According to a 2008 report of UN-HABITAT (2008), the estimated proportion is expected to reach 60% (4.9 billion) by 2030. In Asia, the Asian Development Bank (Evans 2002) predicted that, by 2020, the proportion of urban dwellers will rise to 50%, with city population reaching two billion. It further estimated that there would be more than 150 cities in Asia with populations over one million, and that eighteen of the twenty-seven world megacities (i.e., cities with over 10 million people) will be in Asia.

In general, people and businesses move to cities due to more abundant economic opportunities and improved transportation, education, and housing conditions. Unfortunately, associated with the resultant increased economic activities and growth are many other adverse effects of urbanization that many Asian cities have experienced.

By and large, the most common and direct cause of urbanization in Asia in the last fifty years or so is due to migration of rural populations to the city, especially in the developing countries of East Asia, including China, India, and most Southeast Asian nations. In most developing countries of Asia, owing to the relatively slow economic development and much lower incomes in rural areas, there has been a continuous influx of rural populations to cities seeking jobs and economic opportunities. The generally low educational level and job skills of rural migrants mean that they would have to take up the lowest paid jobs in the cities. Their low income does not allow them to fit into high-cost urban social and living environments. On the other hand, affordable low-cost housing is often insufficient or unavailable to meet the demand of the large number of rural migrants. This has resulted in the development of slums in many cities in the developing world of Asia. Based on the current trend of development, it has been estimated that, in 2020, about half of the city population in Asia will be poor people residing in slums (Evans 2002).

While urban transport and logistics are a vital element of the social and economic activities of a city, they also have created a host of environmental and health problems. Traffic congestion has become a way of life in many Asian cities. As the economic activities of a city escalate, motorization helps to relieve development concentration within the city center by making suburbanization feasible through enhancing mobility, connectivity, and accessibility. On the other hand, these enhancements also cause traffic congestion in the city center due to overwhelming inflow toward the city center. The major cities of Asia show opposite extremes in their performances in urban development and traffic management. Singapore has earned a widely acknowledged reputation as a world model city. Hong Kong and cities in Japan and Korea resemble Singapore with appropriate city planning and well-managed public transportation systems. In the other extreme, Bangkok, Manila,

Jakarta, and many Asian cities are among the most congested cities in the world, including most South Asian cities, which are characterized by narrow streets, heavy traffic, visible haze, and unplanned city architecture (Mistry et al. 2004). These cities have not systematically implemented a cohesive city development plan due to either lack of resources or poor management. Some have applied new traffic management schemes, but received only very limited success because they were implemented in a piecemeal fashion uncoordinated by any consistent long-term strategy (Mills 2000).

Air pollution caused by urban traffic poses undesirable health risks to urban inhabitants. It is responsible for deaths from cardiopulmonary causes, and it increases the risk of nonallergic respiratory symptoms and disease. The pollutants of concern to health are nitrogen dioxide, carbon monoxide, particulate matter (PM) 10 (PM with a diameter of less than 10 μm) and PM 2.5, back smoke, benzene, polycyclic aromatic hydrocarbons (PAHs) and metals, including lead (Krzyzanowski 2005). Kim (2002) concluded from a study of air quality in twenty major Asian cities that, with few exceptions, the chemical, smoke, and particulate matter pollution dramatically exceeded World Health Organization (WHO) limits. The twenty cities studied were Bangkok; Beijing; Busan; Colombo, Sri Lanka; Dhaka; Hanoi; Ho Chi Minh City; Hong Kong; Jakarta; Kathmandu; Kolkata; New Delhi; Mumbai; Manila; Seoul; Shanghai; Singapore; Surabaya; Taipei; and Tokyo. With daily levels of air pollution routinely at levels much higher than the WHO air quality guidelines, and the high population density in these Asian cities, it is certainly a major cause for concern about the impact on the health of Asian populations.

The traffic-generated air pollution problem in the cities of developing countries in Asia is the multiplicative effect of many factors. In a typical situation, the national economy is dominated by few major cities, which continue to attract more business and migration of rural population. Lagging residential and transport infrastructure could not cope with the demand. Coupled with a lack of proper urban land use planning, this has resulted in severe traffic gridlock in the inadequate road network. The problem is further aggravated by the relatively high polluting emission of the traffic caused by the following contributing factors: outdated or ineffective vehicle emission control policy, weak enforcement of vehicle emission control, large percentage of old vehicles in the traffic stream, lack of effective vehicle maintenance and monitoring systems, use of poor-quality fuels, and inefficient public transport systems leading to the partial reliance on low-capacity, environmentally unfriendly forms of paratransit. Figure 2.1 provides a schematic representation of the relationship of the various factors described.

```
┌─────────────────────────────────┐    ┌─────────────────────────────────┐
│         Urban Planning          │    │     Transport Infrastructure    │
│  • Poor land use planning       │    │  • Inadequate investment        │
│  • Uncoordinated land use       │    │  • Poorly planned and designed  │
│    development                  │    │    road network                 │
│  • Overcrowded urban areas      │    │  • Poorly maintained roads      │
│  • Uncontrolled urbanization    │    │  • Outdated and inefficient     │
│                                 │    │    public transit sytems        │
└─────────────────────────────────┘    └─────────────────────────────────┘
```

- Severe traffic congestion and long travel delays
- Highly polluting vehicle emission
- Abundance of low-capacity low-fuel-efficiency transport modes
- Inefficient city logistics operations

```
┌─────────────────────────────────┐    ┌─────────────────────────────────┐
│  Transport Planning and Policy  │    │       Traffic Management        │
│  • Uncontrolled motorization    │    │  • Weak enforcement of traffic  │
│  • Outdated vehicle emission    │    │    regulations                  │
│    standards                    │    │  • Ineffective congestion       │
│  • Use of polluting fuels       │    │    mitigation policy and        │
│  • Inadequate vehicle inspection│    │    measures                     │
│    and enforcement schemes      │    │  • Outdated traffic control     │
│                                 │    │    devices and tools            │
│                                 │    │  • Lack of priority schemes for │
│                                 │    │    public transit systems       │
└─────────────────────────────────┘    └─────────────────────────────────┘
```

FIGURE 2.1 Traffic-related problems in urban systems.

2.2.2 Characteristics of Urban Freight Transport and Logistics

A unique aspect of freight traffic in city logistics is the high-volume movements of low-value consumer packaged goods (CPG), also termed as fast moving consumer goods (FMCG) (Bronnenberg, Dhar, and Dubé 2007). Consumer packaged goods are products that are consumed in daily life, such as personal care products, household care, packaged foods, and beverages, as summarized in Table 2.1. Owing to the large demand for CPG by urban populations, the industry is a major industrial sector in all Asian countries, especially among the end-user industries. For instance, as shown in Figure 2.2, the CPG sector in India has the second highest revenue share among the end-user industries, second only to the metal industry (Natarajan 2008).

TABLE 2.1 Types of Consumer Packaged Goods

Category	Products
Personal care	Medicine and health care products, cosmetics, toiletries, hair care, bath and sanitary, baby and children products, etc.
Household care	Kitchen care, home cleaning, laundry, furniture care, etc.
Office products	Office stationery, furniture care, pantry supplies, office cleaning, janitorial, etc.
Packaged foods	Staple foods, vegetables, meats, fruits, dairy products, bakery products, snack food, etc.
Beverages	Wines, alcohol, juices, soft drinks, health beverages, etc.
Miscellaneous	Newspapers, books, decorative products, branded commodities, etc.

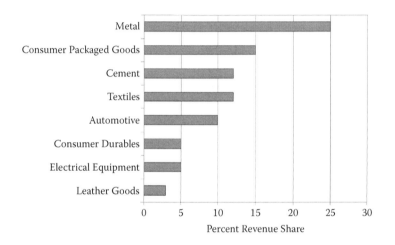

FIGURE 2.2 Revenue shares of end-user industries in India (After Natarajan, 2008).

Consumer packaged goods are typically retail products for end consumers and are used frequently. They are used directly mostly by the end consumers daily, weekly, or regularly over a varying period of time. As such, collection and delivery has to be made frequently at regular intervals over the entire distribution network covering all parts of a city. A typical outbound supply chain involves a number of intermediaries, as shown in Figure 2.3. The corresponding logistics cost consists of four main components: transportation cost, warehousing cost, freight forwarding cost, and miscellaneous cost. Of these, transportation is the largest cost item for the consumer packaged goods sector in Asian countries. For instance, in China, transportation cost accounts for about 55%

FIGURE 2.3 Typical outbound supply chain for logistics of consumer packaged goods.

of the total logistics cost; in India, it is about 60% (Frost & Sullivan Research Service 2006a, 2006b, 2006c, 2006d).

As there are typically a very large number of retail outlets distributed throughout a city, the transportation of consumer packaged goods is characterized by daily trips to multiple destinations, involving many deliveries that tend to be made in small loads. Besides consumer packaged goods, there are other urban freight movements that exhibit similar transportation characteristics. These include collection and delivery of mail, courier services, petroleum tank truck delivery of petroleum to petrol kiosks, and home delivery of merchandise. Though not of the same nature economically, rubbish collection services from private residences and commercial buildings produce a similar impact on road use and traffic as that of other freight transportation operations just described.

Urban freight transport can be classified as external and internal freight movements (D'Este 2000; Ramokgopa 2004). The freight transport operations described in the preceding paragraph are classified as internal freight movements, or intracity freight movements. They are movements that take place within the city boundaries. External freight transport refers to those freight movements in which goods are transported across city boundaries into the city, out of the city, or through the city.

Internal freight transport is dominated by road transport in practically all Asian cities, although there are instances where water transport, pipelines, or rails are used for specific applications. This is because internal urban freight transport deals primarily with the distribution of goods at the end of supply, with multiple destinations and a relatively small delivery load at each destination. This requires a distribution infrastructure with good accessibility and door delivery at each destination. Road transport that could take advantage of the street network of a city is best suited for this purpose.

External freight transport may involve long-haul transport bound for the city. Long-haul trucks are generally relatively large in size and capacity, often in the form of tractor-trailer combination trucks, and are not suitable for door-by-door delivery of goods in the city centers. Truck terminals or distribution centers are usually constructed at designated points of the city to receive these trucks. This arrangement helps to reduce truck traffic in the city center. External freight could also be transported by rail. Intermodal freight transport facilities are necessary for pickup by trucks for delivery to various destinations in the city. The same arrangement is required for freight arriving by air and sea at the airport and seaport, respectively, that serve the city.

The daily time patterns of movements of urban freight traffic have been found to be quite different from the usual patterns observed for urban passenger car traffic. Research studies have consistently shown that the daily peaking characteristics of freight traffic varies with freight sector type and are different from the morning and evening peaks of passenger car traffic. It is likely that the daily freight traffic flow pattern will vary from country to country, or even from city to city, depending on local culture, business practice, and operational constraints. D'Este (2000) reported that, in Australia, intracity urban freight traffic reaches its peak in late morning and early to midafternoon, while external large vehicle trips may have additional peaks in the early morning and later evening. In Singapore, Fwa, Ang, and Goh (1996) and Fwa, Hew, and Teo (2008) reported that truck traffic picked up soon after 4:00 p.m. and peaked between 5:00 a.m. and 7:00 a.m. It plunged after 7:00 a.m. after the passenger car traffic peaked after 7:00 a.m., before picking up again at around 10:00 a.m. until 12:00 noon. The peak at around noontime was lower than the early morning peak. Truck volume dropped slightly after noontime, remaining more or less constant before falling off steadily after 5:00 p.m.

2.3 IMPACTS OF TRANSPORT AND LOGISTICS ON CITY DEVELOPMENT IN ASIA

Today, a quarter of all world exports come from East Asia (Stone 2008). The rapid industrialization has fueled the high rate of population growth in Asian cities. This in turn has generated increased demand for road vehicle-dependent collection and delivery of food products, consumer goods, household and office supplies, mail, petroleum, and urban infrastructure-related construction materials, maintenance, and services supplies. Together with city dwellers' high demand for commuting and intracity mobility, urban freight transportation has placed tremendous pressure on the urban transport systems. Both passenger and freight transportation in many Asian cities have become a bottleneck in urban economic development. This section highlights the adverse impacts of urban freight movements and logistics operations in a city.

2.3.1 Travel Safety

The services provided by freight transport are essential for the efficiency and convenience of modern urban life. However, the presence of freight vehicles is often not received positively by the general public—the motorists in particular. Because of their larger size and less flexible maneuvering capability on the road, trucks are seen by most motorists to be a potential travel safety hazard. The normal traveling speed of trucks is usually slower than passenger cars by 10 km/h or more on normal roads. On urban expressways, speed differences of as much as 20 km/h or more have been recorded (Fwa et al. 2008), and the speed difference is much higher when on upgrades. A truck will take a much longer time than a passenger car to reach a desired speed from its stopping position, and it requires a much longer distance to come to a complete stop from a given speed. In addition, trucks require a much larger radius of curvature for making a turn and bigger space for making U-turns. As a result, freedom and safety of traffic operation on normal roads (two-lane undivided especially) are adversely affected by the presence of heavily loaded trucks. Loading and unloading of goods by trucks at curbside could severely disruption traffic flow as it effectively reduces the number of useable traffic lanes.

The differences in operational performance of trucks and passenger cars create undesirable driving conflicts in a mixed traffic stream. Such conflicts could lead to accidents due to deficiency in road geometric design and unsafe driving of either the truck driver or passenger car driver. Accidents involving heavy trucks and passenger cars are a major concern because of the severity of injuries. A common factor in collisions involving trucks and other vehicles is car drivers' or motorcyclists' ignorance of a truck's performance limitations, such as limitations associated with braking and stopping distance, acceleration and moving upslope, and visibility.

2.3.2 Social Impacts

City logistics operations contribute to traffic congestion directly through urban freight transport and indirectly through generated logistics-related traffic. Most drivers and road users do not welcome heavy trucks on the road. In addition, loading and unloading of goods as well as delivery activities, especially those performed curbside, that create obstruction to traffic and pedestrian movements are seen to be socially disruptive or even threatening the safety of the public. It is ironical that while the public residents are customers who create the demand for movement and transport of goods, they do not in general show much tolerance for the disruption resulting from satisfying such demands.

The general perception of goods vehicles and heavy trucks as nuisances on the road is caused by a number of factors related to the operational characteristics of these vehicles. Being heavier in weight and bulkier in size than passenger cars, goods vehicles and heavy trucks often appear intimidating physically, move much more slowly and tend to delay other traffic in city streets and narrow lanes where overtaking is difficult. In expressways where the slower goods vehicles and heavy trucks usually are restricted within the slow lanes, their presence makes weaving more difficult for traffic entering or exiting the expressways. The higher noise and vibration created by the operations and movements of goods vehicles and heavy trucks are socially intrusive and add to their negative public image. Another major social problem associated with goods vehicles and heavy trucks is the seriousness of road accidents involving these vehicles.

2.3.3 Environmental Impacts

The key environmental issues associated with goods vehicle and heavy truck operation are air pollution, energy consumption, carbon creation, noise, and safety. Except for a few developed countries, air pollution is a problem of concern in most major cities in Asia. It is particularly serious in rapidly urbanizing cities of South, Southeast, and East Asia, including most cities in India and China. In these cities, air pollution levels often exceed WHO levels by a big margin.

Air pollution caused by traffic using fossil fuels is a leading cause of the problem. Goods vehicles and heavy trucks, many of which run on diesel fuel, are known to be a major factor of the pollution problem. For instance, Sperling and Gordon (2008) reported that, in Hong Kong, diesel commercial vehicles produced 90% of particulate matter and 80% of nitrogen oxide emissions from the road sector. Environmental diesel emissions are known to contribute to lung cancer and have been classified as probably carcinogenic to humans (IARC 1988). In Bangkok, traffic policemen are exposed to high levels of automobile-derived particulate air pollution and demonstrate an increased prevalence of respiratory symptoms (Tamura et al. 2003).

Another related environmental concern is the climate change issue on the contribution by transportation to global warming. Studies have provided convincing evidence that carbon emissions have caused global warming, which in turn leads to extreme climate fluctuations and rising sea levels (Moriguchi 2009). Asian countries feature prominently among those that produce the greatest quantity of total CO_2 emission in the world. Figure 2.4 shows that China, India, Japan, and Korea are among the world's top ten CO_2 emission nations. It is also noted that the ten ASEAN (Association of Southeast Asian Nations) member countries together produced more CO_2 emissions than the fifth ranked country, Germany. Statistics from different countries indicate that transportation

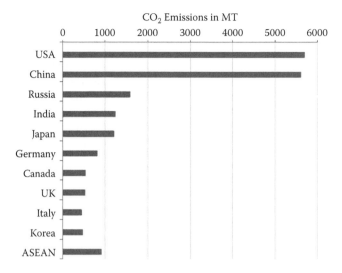

FIGURE 2.4 CO_2 emissions of top ten nations and ASEAN. (After IEA, 2008, http://www.iea.org/textbase/stats/index.asp.)

typically is responsible for about 20% of the total CO_2 emission, second only to industry-generated CO_2 (Jiang 2009; Moriguchi 2009; Shrestha 2009; Shukla 2009).

The key to reducing CO_2 emission of goods vehicles and trucks lies with the type of fuel used. Fossil fuels are known to be the main source of transport-generated CO_2 emission. There is a need to study how the use of fossil fuels for freight transportation could be minimized, and how the use of environmentally friendly, low-carbon or zero-carbon technologies and alternative fuels could be implemented or enforced.

2.4 NEED FOR INCORPORATING CITY LOGISTICS INTO TRANSPORT AND LAND USE PLANNING

Because of the importance of trade and logistics to the economy of a city, ideally, transport and land use planning must accommodate the needs of trade and logistics in the city master development plan. Unfortunately, few cities, if any, have these needs included in their master development plan. The traditional transport and land use planning concept and methodology do not specifically address the needs for city logistics operations. This is because the characteristics of urban goods transport are very different from those of passenger transport and the transport plan that is drawn up based on passenger transport would not be able to satisfy the requirements of goods transport. According to studies on urban goods transport planning (e.g., Routhier 2008; Debauche 2008), the following characteristics of urban transport logistics have to be considered in land use planning:

a. City logistics consists of a wide variety of stakeholders, goods vehicles, and organizations.
b. There is a large diversity of trip chain patterns in urban goods transport.
c. Urban goods transport operates within integrated supply chain management while passenger transport serves individual needs.
d. Vehicular maneuvering capability and travel speed of goods vehicles are very different from those used for passengers, and additional consideration in space and road geometric requirements is necessary.
e. Urban freight operations, including loading and unloading and space requirements for distribution and storage, exert stronger impacts on land use development than individual passenger transport needs.
f. The urban logistics chain of operations does not stop at the city boundary. Freight activities typically extend far beyond city boundaries, involving transfer of goods from regional distribution centers and long-distance supply transportation.

In traditional urban planning, passenger transport has received the top priority. The effect of freight traffic is often considered by converting heavy vehicles into equivalent passenger car units. Goods transport in North American and European cities represents from 10% to 18% of urban traffic (Stantchev and Whiteing 2006) and 20% to 25% in terms of car equivalent vehicle-kilometers (Debauche 2008). In Singapore, the volume of truck traffic varied mostly between 10% and 20% of total flow volume (Fwa et al. 1996, 2008). In developing cities in Asia, the situation can be quite different. Due to lack of a well-developed city logistics system, a substantial proportion of goods supply and delivery services is still made in old-fashioned manners through the use of lorries, converted vans, and even motorcycles.

This is clearly inadequate because the impact of city logistics on urban planning is far greater than its effect on traffic flow, as indicated by the characteristics listed previously. In this regard, it is appropriate to stress the need to enhance traditional transport and land use planning by incorporating the operational characteristics and requirements of city logistics. In addition to the conventional household survey for data collection of passenger car trips, it is also necessary to collect data concerning city logistics supply chain behavior and land use requirements. There is also a need to establish the relationship between city logistics supply characteristics and the retailing characteristics, as well as the downstream customer purchasing behavior. Last but not least, an integrated approach of urban freight traffic and passenger car traffic modeling has to be developed for comprehensive urban land use planning.

Because of the lack of consideration for urban logistics needs in land use planning in many cities, urban freight operations are often seen to be disruptive to urban traffic movements and conduct of business. As a result, an urban planning authority could only react to urban traffic problems by imposing restrictive measures to freight traffic and city logistics operations. The following forms of restriction are commonly found in cities of Asia:

 a. Access restriction of freight traffic within specified time periods of the day
 b. Access restriction of freight traffic to specified areas of the city
 c. Restriction on the size of freight vehicles within the city
 d. Restriction on the weight of freight vehicles within the city
 e. Restriction on the modes of loading and unloading within the city
 f. Restriction on parking of freight vehicles within the city
 g. Restriction on travel speed of freight vehicles

The need for integrated urban land use planning incorporating freight and city logistics requirements is even more apparent in the case of hub port cities. Port activities and freight flows present a major consideration in urban development to virtually all coastal and inland port cities. In the last two decades or so, with accelerated logistics integration due to globalization and advanced technical advances in transportation engineering, a new pattern of freight distribution has emerged and the importance of a port in terms of its functional role in a city's economic development is growing with time. This requires corresponding adjustments in land use planning to accommodate such changes.

The case of Singapore is an excellent example. Prior to 1965, in the preindependence days, Singapore's maritime trading activities started along the Singapore River, right in the heart of the city center, and Keppel Harbor, which is located at the southern shoreline of the city. The first postindependence land use master plan of Singapore, issued in 1971, featured long-term strategic plans for the southern coastal areas. Planned port expansion has been implemented systematically in the last thirty years, moving the main port terminal activities toward the western industrial areas away from the city center. Land acquisition and physical development in the coastal areas have taken place to accommodate the rapid increase in port container traffic and the land freight traffic between port terminals and the western industrial zone. While the 2008 container throughput of the Port of Singapore reached 27.9 million TEU (20-foot equivalent units), the port land use and development plan is already in place for the design throughput capacity of 50 million TEU.

Poor transport and land use planning or implementation cause economic losses and impede sustainable growth of a city. Unlike Singapore and other cities in developed Asian countries such as Japan and Korea, most developing Asian cities are today severely affected by daily congestion caused by passenger commuting traffic due to inadequate transport and land use planning. For instance, a recent study showed that 60% of the time spent in the Jakarta road network is delayed time, meaning that we spend most of our travel time here stuck in traffic congestion. The economic loss attached to congestion in Jakarta is huge; it can reach more than Rp 8.8 trillion (US$755 million) each year (Susantono 2008). Similar urban traffic problems are found in other Asian cities such as Bangkok, Manila, New Delhi, Kuala Lumpur, etc. (Agarwal 2008; Hanaoka 2005). Vergel (2004) reported on studies and various plans that have been attempted to reduce traffic congestion and pollution problems in Manila and recommended that improved transportation and environment simulation models be developed to cover the transportation networks in metro Manila more effectively.

2.5 EMERGENCE OF THE MEGALOPOLIS

In spite of the social issues, the abundance of market opportunities and the benefits of economies of scale have driven the global trend of concentrating key economic functions among major megacities around the world. This trend is further fueled by the globalization of the economy that leads to intense competition among cities around the world for investment capital, trade, knowledge, and technology. In many Asian countries, such megacities are taking shape at an accelerated pace, serving as political and financial centers and economic hubs, as well as transportation and logistics hubs. Examples of such Asian megacities are Tokyo in Japan, Seoul in Korea, Shanghai in China, Jakarta in Indonesia, Bangkok in Thailand, and Manila in the Philippines.

Riding on the previously mentioned trend is the emergence of the "megalopolis," which adds an additional dimension to megacities. Megalopolis refers to a multicity corridor characterized by a complex network of economic centers connected by efficient intercity high-speed transportation links. Gottmann (1961) first used this term to describe the dominance of the U.S. Eastern Seaboard metropolitan areas, represented by the Boston–New York–Philadelphia–Baltimore–Washington, DC corridor of national/international centers for governmental, financial, business, academic, and media services. In Asia, the Tokyo–Nagoya–Osaka corridor in Japan is the first example of a megalopolis (Karan and Gilbreath 2005). The latest examples of megalopolis development in Asia are found in China. The Pearl River Delta on the southern coast of China and the Yangtze River Delta on the eastern coast are currently the fastest growing megalopolises in the world (Shen 2007; Ning 2007). Table 2.2 shows the cities located within each of the three megalopolises of Asia.

A megalopolis produces a multiplicative effect economically by coalescing the consumer markets of contributing cities, and through

TABLE 2.2 Megalopolises of Asia

Megalopolis	Cities and Populations	Total Population
Tokyo–Nagoya–Osaka	Tokyo (12.4 million), Yokohama (3.6 million), Nagoya (2.3 million), Kyoto (1.5 million), Osaka (8.6 million), Kobe (1.5 million)	29.9 million
Pearl River Delta	Hong Kong (7.0 million), Macau (0.5 million), Guangzhou (10.2 million), Shenzhen (9.0 million), Zhuhai (1.2 million), Dongguan (7.6 million), Foshan (3.4 million), Jiangmen (4.0 million), Zhaoqing (3.9 million), Huizhou (3.3 million), Zhongshan (2.5 million)	52.6 million
Yangtze River Delta	Shanghai (18.9 million), Nanjing (6.5 million), Suzhou (5.9 million), Wuxi (4.3 million), Changzhou (3.5 million), Yangzhou (4.6 million), Zhenjiang (2.7 million), Nantong (0.8 million), Taizhou of Jiangsu (5.1 million), Hangzhou (7.9 million), Ningbo (5.4 million), Huzhou (2.5 million), Jiaxing (3.3 million), Shaoxing (4.3 million), Zhoushan (1.0 million), Taizhou of Zhejiang (1.4 million)	78.1 million

Note: Population of each city is estimated for 2008 based on city population data from United Nations Statistics Division (2009). Demographic and social statistics. http://unstats.un.org/unsd/demographic

complementary and coordinated development of multicity commerce and industry. It often functions as a key regional or national economic engine and contributes significantly to the national gross domestic product (GDP) of a country. The dominance of the Tokyo–Nagoya–Osaka megalopolis in the Japanese economy is a clear example. In China in 2005, the Yangtze River Delta region, with 1% of the nation's land and 6% of total population, produced 18.6% of national GDP and absorbed 43.5% of the nation's total utilized foreign direct investment capital (Ning 2007). In the same year, the Yangtze River Delta megalopolis and the Pearl River Delta megalopolis, together with the Beijing–Tianjin–Hebei region, which is another emerging megalopolis in China, contributed a domestic share 42.4% in GDP and 77% in export value of the country (Shen 2007).

It should be emphasized that the formation of a megalopolis is facilitated by the availability of high-speed transportation and communication linkages, which are only possible with the modern day's technological advancements. The development of high-speed expressways and express railway and freight services significantly reduces trip times of intercity movements of people and goods. Business trips and freight services become more frequent as a result, with a tendency toward increasingly coordinated economic development among the cities that further encourages intercity trade and logistics activities.

The emergence of the megalopolis has a major implication on urban and intercity transportation and business and causes significant changes in the economic and industrial structures of the member cities. It in turn necessitates corresponding changes in the logistics network across the region. Relocation of logistics centers and setting up of specialized distribution centers would be necessary. In view of the much larger population size involved in a megalopolis in Asia, as compared to megalopolises in Europe and North America, the scale and intensity of trade and logistics activities is expected to be higher.

A common feature of all megalopolises is the presence of one or more major logistics and/or distribution hubs strategically located within the megalopolis, due to the large population base and active intercity trade. For instance, the Tokyo–Nagoya–Osaka megalopolis is served by the top ports in Japan located at Tokyo, Yokohama, Kobe, Nagoya, and Osaka. The Pearl River Delta megalopolis has Hong Kong Port, Shenzhen Port, and Guangzhou Port—all ranked within the world's top ten busiest container ports for 2008. In the Yangtze River Delta megalopolis, the Shanghai Port and the Ningbo Port were, respectively, the second and seventh ranked container ports in the world in 2008. These ports serve as the gateway to the large market of their respective megalopolises. The development of an efficient and well connected network of road and rail system, or inland waterway system, is crucial to the sustainable growth of the logistics industry of a megalopolis.

2.6 OPPORTUNITIES AND CHALLENGES OF CITY LOGISTICS DEVELOPMENTS IN ASIAN CITIES

2.6.1 Enhancing Transport and Land Use Planning Incorporating City Logistics Needs

A dilemma faced by cities in the developing countries of Asia is that while the cities contribute to a large proportion of a nation's economy and wealth, their progress is also very much hindered by the very economic and trade activities that generate the wealth. Inefficient city logistics operations cause substantial economic loss and impede further trade and city logistics growth. The problem is caused by one or more of the following factors:

- Poor or outdated land use planning that does not take into consideration the operational requirements of city logistics
- Weak implementation or enforcement of land use plans
- Inefficient city logistics infrastructure
- Severe traffic congestion

There is no single simple solution that can be applied for all cities, because each city has its own characteristics and peculiarities. Nevertheless, experiences in cities of developed nations in Asia have shown that with a rational land use plan that incorporates the needs of city logistics, plus a systematic implementation and enforcement strategy of land use development and traffic management, improved traffic conditions and city logistics operations could be achieved.

To enhance the capability and effectiveness in land use planning and implementation, the authorities of Asian cities could look into the possibility of adopting advanced transportation and management technologies available today and best practices in other Asian cities. For instance, information and communication technologies have already been widely used in the logistics and the urban transport sectors. In addition, intelligent transport systems (ITS) such as traffic monitoring systems and vehicle routing software can be applied to enhance both passenger and freight traffic movement. The potential benefits of integrating traffic management systems with city logistics systems must not be overlooked. The development of an urban distribution network infrastructure and strategically located consolidation centers can help to improve deliveries to urban areas.

2.6.2 Exploiting Alternative Transport and Distribution Modes for Freight

In view of the unique social culture and business practices in many Asian cities, the innovative use of available technology or mode of transport and distribution suitable for local conditions can be considered. In addition to the conventional approaches of regulating and controlling urban freight traffic, other approaches such as spatial planning and land management could be employed to achieve better results without resorting to restricting urban logistics operations. In this regard, increased cooperation and dialogue among the stakeholders would be beneficial. The stakeholders include local authorities, logistics operators, transport providers, retailers, and community representatives.

Depending on the geographical location and the characteristics of a logistics supply chain of a city, the possibility of exploiting multimodal city logistics operation must be explored. Road-based distribution systems have been the "obvious" choice for most logistics operators because of its convenience in providing door-to-door service. However, with the increasing congested road network and costlier land space in most cities, the advantages of alternative transport modes or multimodal operations must not be ignored. For instance, the choice of distribution center sites could consider the merits of locations in proximity to railways and waterways. Multimodal freight transport involving non-road-based transport such as rail and water transport offers relief to road congestion, with possible cost and environmental benefits.

For a coastal port with a seaport, cabotage trade in coastal waters would be an attractive means of freight transport in terms of cost and handling that was not necessarily limited to bulk commodities. Similarly, city waterways could be an effective supplementary mode of freight transport to relieve the pressure of urban freight transport within a city. For a city with a good canal and waterway system, the accessibility of a water distribution network can be extensive. Waterways could be utilized for transport construction materials and waste. This would help reduce road use with its associated problems of congestion, accidents, and pollution. Bangkok is one of the Asian cities that could take advantage of its extensive waterway network for this purpose. It used to be known as the "Venice of the East." Unfortunately, since the twentieth century, strong economic development and motorization in the city caused a rapid transformation from water-based transport to road-based transport (Tanaboriboon, Hanaoka, and Iamtrakul 2005).

2.6.3 Achieving Environmentally Friendly City Logistics for Sustainable City Development

While city logistics activities are essential to the economic and social life of a city, they have contributed to traffic congestion and adverse environmental impacts. Adopting good city logistics practices and strategies that do not create adverse environmental impacts is key to the long-term sustainable growth of a city. The main environmental issues created by city logistics operations are air pollution, noise, and carbon creation, as well as traffic safety and congestion that cause social distress. They are of major concern because of their adverse impacts on the sustainable development of a city by causing detrimental effects to the health of citizens and decreasing the quality of the living environment of the community.

Rational land use planning and the policy of systematic coordinated implementation will go a long way to eliminate much of the environmental ills of city logistics. A well-planned land use development would isolate city logistics noise from populated residential and social areas, and main corridors of commuting traffic and social activities, thereby keeping noise and traffic safety problems under control and reducing the influence of the detrimental health effects of air pollution. It is unfortunate that due to the lack of political will and institutional inefficiency, many cities in the developing countries of Asia have not yet been able to embark on a consistent long-term policy to implement those measures that have been proved effective by many other cities in Asia and the world at large.

Energy consumption and the type of fuel used by city freight traffic are another environmental sustainability issue that deserves attention by Asian city authorities. Efficient traffic management that improves traffic flow and optimal fleet schedule and route planning for freight vehicles will reduce energy consumption and help save wasteful consumption of scarce energy resources. Asian cities must also be conscious in limiting the use of fossil fuel for both passenger cars and freight vehicles so as to reduce their overall carbon footprint. They should participate in the ongoing global collaborative effort toward becoming a low-carbon society and take advantage of new technological developments that might become available to achieve the structural transformation for a sustainable economy.

REFERENCES

Agarwal, O. P. (2008). Urban mobility in India: Initiatives and lessons learned. Meeting report, World Urban Transport Leaders Summit 2008. Singapore, November 4–6, 2008.

Bronnenberg, B. J., Dhar, S. K., and Dubé, J. P. (2007). Consumer packaged goods in the United States: National brands, local branding. *Journal of Marketing Research* 44(1): 4–13.

Debauche, W. (2008). Why local decision makers are reluctant to urban freight modeling and data collection. BESTUFS WP3 Rome Roundtable, Rome, Italy.

D'Este, G. (2000). Urban freight movement modeling. In *Handbook of transport modeling,* ed. D. A. Hensher and K. Joh, chap. 33. Bingley, England: Emerald Group Publishing.

Evans, W. (2002). Strategy to meet challenges of Asia's megacities. News release no. 075/02, 35th Annual Meeting of the Board of Governors, Asia Development Bank.

Frost & Sullivan Research Service (2006a). Strategic analysis of 3PL markets in the Chinese FMCG sector. Frost & Sullivan, Singapore.

——— (2006b). Strategic analysis of 3PL markets in the Indian FMCG sector. Frost & Sullivan, Singapore.

——— (2006c). Strategic analysis of 3PL markets in the Malaysian FMCG sector. Frost & Sullivan, Singapore.

——— (2006d). Strategic analysis of 3PL markets in the Indonesian FMCG sector. Frost & Sullivan, Singapore.

Fwa, T. F., Ang, B. W., and Goh, T. N. (1996). Characteristics of truck traffic in Singapore. *Journal of Advanced Transportation* 30 (2): 25–46.

Fwa, T. F., Hew, G. H., and Teo, Z. B. (2008). Study of truck traffic around Port of Singapore. *Proceedings 2008 Annual Symposium of Pavement Technology,* May 23, 2008, Singapore.

Gottmann, J. (1961). *Megalopolis: The urbanized northeastern seaboard of the United States.* New York: Twentieth Century Fund.

Hanaoka, S. (2005). Urban transport problems, issues and some policies: A case of Bangkok. Special session, 6th EASTS Conference, Eastern Asia Society for Transportation Studies, Bangkok, September 22, 2005.

IARC (1988). IARC monographs on the evaluation of carcinogenic risks to humans, vol. 46. Diesel and gasoline engine exhausts and some nitroarenes. Lyon, France: WHO.

IEA (2008). IEA statistics. International Energy Agency. http://www.iea.org/textbase/stats/index.asp (accessed April 18, 2009).

Jiang, K. J. (2009). China low carbon society scenarios. Presented at Symposium on Path toward Low-Carbon Society: Japan and Asia, February 12, 2009, Tokyo. Ministry of the Environment, Japan, and National Institute for Environmental Studies (NIES), Japan.

Karan, P. P., and Gilbreath, D. (2005). *Japan in the 21st century: Environment, economy, and society.* Lexington: University Press of Kentucky.

Kim, C. (2002). Air pollution in the megacities of Asia project. Presented at concluding workshop: Reducing Vehicle Emissions Project, February 28–March 1, 2002, ADB Auditorium (zones A–C). Manila, the Philippines.

Krzyzanowski, M. (2005). Health effects of transport-related air pollution. World Health Organization.

Mills, E. S. (2000). The importance of large urban areas and government roles. In *Local dynamics in an era of globalization,* ed. S. Yusuf, W. P. Wu, and S. Evenett. Oxford, England: Oxford University Press.

Mistry, R., Wickramasingha, N., Ogston, S., Singh, M., Devasiri, V., and Mukhopadhyay, S. (2004). Wheeze and urban variation in South Asia. *European Journal of Pediatrics* 163 (3): 145–147.

Moriguchi, Y. (2009). Transportation in a low-carbon society. Presented at Symposium on Path toward Low-Carbon Society: Japan and Asia, February 12, 2009, Tokyo. Ministry of the Environment, Japan, and National Institute for Environmental Studies (NIES), Japan.

Natarajan, K. K. (2008). Survey of domestic trucking transport industry of India. Master of science project report, Department of Civil Engineering, National University of Singapore.

Ning, Y. M. (2007). Yangtze Delta region: Urbanization development and megalopolis restructuring. Presented at 4th International Conference on Population Geographies, Chinese University of Hong Kong, July 10–13, 2007, Hong Kong.

Ramokgopa, L. N. (2004). City logistics: Changing how we supply. *Proceedings 23rd Southern African Transport Conference,* Pretoria, South Africa, July 12–15, 694–702.

Routhier, J. L. (2008). Integration of urban goods movement into the whole urban transport system. BESTUFS WP3 Rome Roundtable, Rome, Italy.

Shen, B. (2007). Megalopolis development and governance in China. Working paper, Institute for Spatial Planning & Regional Economy, National Development & Reform Commission, China.

Shrestha, R. M. (2009). Thailand low carbon society scenarios. Presented at Symposium on Path toward Low-Carbon Society: Japan and Asia, February 12, 2009, Tokyo. Ministry of the Environment, Japan, and National Institute for Environmental Studies (NIES), Japan.

Shukla, P. R. (2009). India low carbon society scenarios. Presented at Symposium on Path toward Low-Carbon Society: Japan and Asia, February 12, 2009, Tokyo. Ministry of the Environment, Japan, and National Institute for Environmental Studies (NIES), Japan.

Sperling, D., and Gordon, D. (2008). Two billion cars transforming a culture. *TR News* 259: 3–9. Transportation Research Board.

Stantchev, D., and Whiteing, T. (2006). Urban freight transport and logistics—An overview of European research and policy. Directorate-General for Energy and Transport, European Commission.

Stone, S. F. (2008). Asia's infrastructure challenges: Issues of institutional capacity. ADB Institute working paper no. 126, Asian Development Bank Institute.

Susantono, B. (2008). Stuck in Jakarta traffic: Is the gridlock coming early? *Jakarta Post,* December 12, 2008.

Tamura, K., Jinsart, W., Yano, E., Karita, K., and Boudoung, D. (2003). Particulate air pollution and chronic respiratory symptoms among traffic policemen in Bangkok. *Archives of Environment Health* 58: 201–207.

Tanaboriboon, Y., Hanaoka, S., and Iamtrakul, P. (2005). Water transport in Bangkok. Third International Symposium on Southeast Asian Water Environment Program, University of Tokyo, Bangkok, Dec. 7, 2005.

UN-HABITAT (United Nations Center for Human Settlements Habitat) (2008). State of the world's cities 2008/2009—Harmonious Cities, Monitoring and Research Division, United Nations.

United Nations (2005). World urbanization prospects: The 2005 revision. Population Division, Department of Economic and Social Affairs, United Nations.

United Nations Statistics Division (2009). Demographic and social statistics. http://unstats.un.org/unsd/demographic (accessed April 4, 2009).

Vergel, K. B. N. (2004). Urban transportation and the environment in metro Manila. Presented at Seminar of Department of Built Environment, Tokyo Institute of Technology, April 2, 2004.

CHAPTER 3

Healthy Transport

Russell G. Thompson

CONTENTS

3.1 Introduction	54
3.1.1 Transport and Health	54
3.1.2 Health Problems	54
3.1.3 Burden of Disease	55
3.1.4 Physical Activity	56
3.1.5 Physical Activity Guidelines	56
3.2 Active Transport	57
3.2.1 Modes	57
3.2.2 Urban Ecology	58
3.2.3 Walkability	58
3.2.4 Self-Change	60
3.2.5 Journey Planning	60
3.3 Measurement	62
3.3.1 Self-Reported Methods	62
3.3.2 Accelerometers	63
3.3.3 Global Position Systems	67
3.4 Modeling	68
3.4.1 Estimating Physical Activity Levels	68
3.4.2 Energy Balance Modeling	70
3.5 Evaluation	70
3.5.1 Benefit-Cost Analysis	70
3.5.2 Multicriteria Methods	71
References	72

3.1 INTRODUCTION

3.1.1 Transport and Health

Most people in the world live in cities and urbanization is predicted to increase dramatically with the growing number of persons moving to cities from rural areas. This presents major challenges for engineers to enhance livability, mobility, and sustainability.

Health impacts of transport are largely related to road safety, vehicle emissions, noise, and physical activity. Road safety initiatives are addressed in Chapter 11. Vehicle emissions and noise are primarily related to vehicle technologies, and improvements in alternative fuels as well as improved engine efficiency are significantly reducing their effect on health. This chapter examines how transport can lead to health benefits by increasing physical activity levels.

3.1.2 Health Problems

The lifestyle of persons living in large cities is becoming increasingly sedentary, with a large proportion of persons' wake time spent sitting or exercising little. A number of factors are contributing to the reduced amount of physical activity persons are undertaking compared with previous generations living in cities, such as the following:

a. Mechanization of work and transport
b. Reduced participation in sport and recreation
c. Increased concern for personal security (especially children and the elderly)
d. Increased indoor home entertainment (e.g., television and computers)

Physical inactivity has been linked to a number of common and increasingly prevalent health problems, such as cardiovascular disease and a number of chronic diseases such as cancer (colon and breast), diabetes mellitus, osteoporosis, and depression (Haskell et al. 2007; Bouchard and Shephard 1994; Blaire and Brodney 1999).

It is common for persons who have type 2 diabetes also to have hypertension and be obese. This clustering of risk factors is called the metabolic syndrome (WHO 2006). Lifestyle changes, such as increasing physical activity, have been shown to provide an effective means of preventing diabetes.

Physical activity protects or may protect against cancers of the colon, breast, and endometrium (AICR 2007). Physical activity protects again overweight, weight gain, and obesity. A major recommendation of the cancer report is to "be physically active as part of everyday life."

Governments are concerned about the increasing prevalence of obesity and chronic diseases. Research shows that approximately 3.3 million Australians are obese, with another 5.6 million overweight (Australian Institute of Health and Welfare 2003). In the state of Victoria, obesity is ranked second as a cause of premature death and disability, contributing to 8% of the overall burden of disease (DHS 2005). Disturbingly, the prevalence of overweight Australian children almost doubled during the last decade, while levels of obesity more than tripled. Conservative estimates indicate that 23% of Australian children are overweight or obese (Booth et al. 2001). Over the last two decades obesity rates among adults have also increased. Between 1980 and 2000, rates of obesity or overweight among males aged twenty-five to sixty-four years increased from 47.3% to 65.7%. Among women, the rate increased from 27.2% in 1980 to 46.5% in 2000 (DHS 2005). Due to the low rates of activity levels in Australia, it is suggested that no one group is more "at risk" from inactivity and that the whole community should be targeted, as there is a need to increase physical activity levels across most groups.

By improving physical activity through active transport, the development of chronic diseases like diabetes and cardiovascular disease can be prevented, with significant personal, social, and economic benefits. There is strengthening evidence around the benefits of physical activity for the community as a whole. The National Public Health Partnership update on evidence for physical activity and health, Getting Australia Active II (Bull et al. 2004), quotes a recent Danish study that showed cycling to work reduces mortality risk, thus providing clear and positive evidence regarding active commuting.

3.1.3 Burden of Disease

The effect of diseases within communities can be analyzed by investigating the amount and quality of life lost. The disability adjusted life year (DALY) incorporates both the number of years of life lost (YLL) from premature mortality as well as the equivalent number of "healthy" years of life lost as a result of disability (YLD) (Murray 1994; Murray and Acharya 1997). A number of risk factors have been identified for Victorians, with obesity and physical inactivity accounting for 8% and 4.1% of the overall burden of disease respectively (DHS 2005).

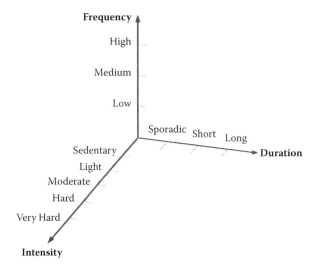

FIGURE 3.1 Physical activity bouts.

3.1.4 Physical Activity

There are four common types of broad categories of activities where persons undertake physical activity:

a. Sport and recreation
b. Work
c. Household
d. Transport

Bouts of physical activity are generally represented by frequency, intensity, and duration (Figure 3.1). There are defined levels for the categories in each dimension (Esliger and Tremblay 2007). Over the course of a day, physical activity can also be characterized by when it occurs, time of day (temporal), and where it occurs (spatial; indoors or outdoors).

3.1.5 Physical Activity Guidelines

Many countries have developed health guidelines that provide recommendations regarding the desirable levels of physical activity. The U.S. guidelines (U.S. Department of Health and Human Services 2008) recommend:

> For substantial health benefits, adults should do at least 150 minutes (2 hours and 30 minutes) a week of moderate-intensity, or

75 minutes (1 hour and 15 minutes) a week of vigorous-intensity aerobic physical activity, or an equivalent combination of moderate- and vigorous-intensity aerobic activity. Aerobic activity should be performed in episodes of at least 10 minutes, and preferably, it should be spread throughout the week.

The current Australian physical activity guidelines are to "put together at least thirty minutes of moderate-intensity physical activity on most, preferably all, days." Recently released Irish guidelines for adults (aged eighteen–sixty-four) called for "at least 30 minutes a day of moderate activity on 5 days a week (or 150 minutes a week)" (Department of Health and Children, Health Service Executive 2009). Health guidelines can generally be achieved with only a moderate amount of cycling and walking per day incorporated into the journey to work (Shephard 2008).

3.2 ACTIVE TRANSPORT

3.2.1 Modes

The term "active transport" relates to

> ...physical activity undertaken as a means of transport. This includes travel by foot, bicycle and other non-motorized vehicles. Use of public transport is also included in the definition as it often involves some walking or cycling to pick-up and from drop-off points. Active transport does not include walking, cycling or other physical activity that is undertaken for recreation. (NPHP 2001)

Therefore, increases in active transport are likely to have significant direct health benefits. Indirect health benefits may also accrue from reduced environmental pollution and increased community cohesion through increasing physical activity and use of public transport or by walking or cycling.

Transport is a major type and common form of physical activity. Active transport (AT) is any form of transport that is human powered (Healthy Living Unit 1993). Common forms of AT include nonleisure walking and cycling (Berrigan et al. 2006). Li and Rissel (2008) indicated that people who used active transport for their journey to work were the least likely to be overweight or obese after taking into account leisure-time physical activity.

Walking is the most common form of physical activity. It is accessible to a large proportion of the population. Walking for transport is growing for journey-to-work trips in Australia but has declined in the last generation for school children.

Cycling is accessible to a wide age group, environmentally friendly, and a fast means of transport in urban areas. Cycling is also an effective means of undertaking physical activity. It can be undertaken by a wide variety of ages and is affordable.

However, a number of barriers, including environmental factors and safety concerns, have been identified that are limiting greater participation in cycling in Australian cities (Bauman et al. 2008).

3.2.2 Urban Ecology

A main premise in ecological systems theory is that "human behavior cannot be understood without taking into consideration the context in which a person is embedded" (Davison and Campbell 2005). An ecological model of physical activity was recently used to identify a range of individual, social environment, physical environment, and public and regulatory factors that influence participation in cycling in Australia (Bauman et al. 2008).

Characteristics of local areas, including urban density, street design, type of housing, proximity of bike paths, and access to community facilities, have been shown to influence participation rates of bicycle riding significantly (Newman and Kenworthy 1999; Handy 2004). The physical environment's influence on cycling and health largely relates to urban design (especially density and land-use mix) and the provision and accessibility of bicycle infrastructure (Bauman et al. 2008).

Salmon and King (2005) introduced a social ecological model of physical activity that recognizes the influences of individual, sociocultural, and environmental policy on physical activity behavior for general and child-specific populations. They evaluated approaches that have been applied among the population of children and adolescents, including school-based, family-based, primary care, community-based, and active transport approaches. Their evidence showed that to increase physical activity and reduce sedentary behavior, there is a need to apply the full social ecological model in the interventions and a combination of approaches.

3.2.3 Walkability

Environments can either support or restrict healthy behavior. Individual, social/cultural, environmental, and policy factors all influence physical activity participation (Sallis, Bauman, and Pratt 1998). A number of neighborhood factors have been identified that promote walking, including:

- Residential density
- Street connectivity

- Land use mix
- Retail facilities

A walkability index has been defined that provides a composite measure of the local physical environment's features that have been shown to influence positively the amount of walking undertaken (Frank et al. 2006). This is determined by adding the scores of net residential density (residential units divided by acres in residential use), street connectivity (intersections per square kilometer), land-use mix (a function of a range of education, entertainment, residential, retail, and office uses), and retail floor area (retail building floor area divided by retail land area). Street connectivity is given twice the weighting of the other variables.

Increases in this index were shown to be associated positively with time spent in physically active travel and correlated negatively to body mass index and vehicle miles traveled, as well as emissions generated in King County, Washington. A study conducted in Adelaide, Australia, also confirmed strong positive association with weekly frequency of walking for transport and the walkability index (Owen et al. 2007).

A number of other factors have been shown to influence the amount of walking undertaken in local areas (VicLANES 2007), including:

- Functionality (e.g., number of street segment sides with hard pavement surface)
- Safety (e.g., number of segments with crossings)
- Aesthetics (e.g., garden maintenance)
- Destinations (e.g., number of recreational facilities within a 2 km radius)

According to a recent Australian Bureau of Statistics (ABS) publication, in 2006, proximity of home to place of work or study (59%) and exercise and health (49%) were the two most important reasons why people usually walked or cycled. Distance involved is the one significant reason why people did not usually walk or cycle to their place of work or study (70% in 2006) and this was reported by people in each age group.

Civil engineers have a major role in developing a safe and convenient environment for walkers and cyclists. Engineers also have an important role in monitoring walking and cycling and evaluating new facilities. Through improved urban design and walking and bicycle infrastructure, transport engineers and planners can promote higher levels of active transport in cities. Engineering is required to provide lanes, paths, and end-of-journey facilities for cyclists such as showers, parking, and lockers (AUSTROADS 1999). Walkers require safe walkways and paths requiring detailed design and construction of treatments for crossing roads and providing access to public transport and worksites (AUSTROADS 1995).

3.2.4 Self-Change

Approaches based on psychology aimed at changing bad habits, addictions, or unhealthy behavior (e.g., smoking and alcohol abuse) can be applied to assist persons increase the amount of participation in active transport. A five-stage model has been developed to represent the stages of self-change (Prochaska and DiClemente 1983; Prochaska, Norcross, and DiClemente 1994):

a. Precontemplation
b. Contemplation
c. Preparation
d. Action
e. Maintenance

At the precontemplation stage, persons do not see their problem and hence there is the need to raise awareness of the problem and promote the need for change. At the contemplation stage, the problem is acknowledged but the potential benefits of change need to be recognized. Persons in the planning stage are planning to take action within the next month. Persons often require more details and need to make small progressive steps. Persons taking action have modified their behavior and surroundings. In the maintenance stage, there is a need to consolidate the gains attained and avoid lapses and relapses. Social and peer support can be important. The termination is reached when the problem no longer presents any temptation.

Information plays a key role in the change process. This involves presenting persons with details of safe routes and the benefits of walking or cycling.

3.2.5 Journey Planning

Journey planners provide an effective means of disseminating information about a variety of public transport options to travelers for specific trips (www.metlinkmelbourne.com.au). An active transport journey planner has been developed that provides travelers with a ranking of active transport options by incorporating multiple objectives

(Hu, Thompson, and Zaman 2008). A case study was developed to test the methodology. The objectives considered were personal energy expenditure, travel time, travel cost, CO_2 emission, and energy resource consumption concerning sustainability. The active transport journey planner was developed in Excel to allow users to set their constraints for most objectives and give their corresponding weightings, respectively (Figures 3.2 and 3.3). The recommended transport solution (the least total disutility one) and ranking of other options along with their detailed objective-related information are derived. Initial results show that this methodology could be applied in selecting more informed transport solutions based on a user's multiobjective preferences. In addition, transport options incorporating more cycling and walking are more likely to deliver healthier and more sustainable solutions to users if social and environmental concerns were considered beyond economic issues.

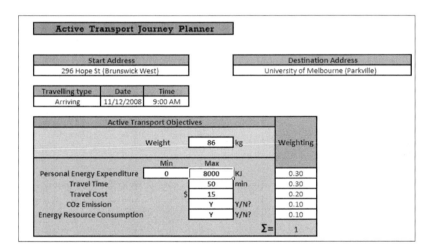

FIGURE 3.2 (See color insert.)   Active transport journey planner (input).

FIGURE 3.3 (See color insert.)   Active transport journey planner (output).

3.3 MEASUREMENT

It is important to be able to estimate the energy expenditure for individuals in order to identify groups that are not meeting health guidelines. However, it is challenging to measure the amount of physical activity undertaken over an extended period accurately (AICR 2007). It is common for questionnaires to be used to provide general information, although these have a number of limitations (Shephard 2003). Accurately estimating the frequency, duration, and intensity levels of walking and cycling is difficult in conventional travel surveys. Recent developments in accelerometers and global position systems (GPS) provide a means of improving the precision associated with estimating physical activity levels.

3.3.1 Self-Reported Methods

Numerous surveys are administered to quantify the amount of physical activity undertaken in daily activities of individuals. Self-reporting is a common way of gaining details. For example, the Active Australia Survey (Australian Institute of Health and Welfare 2003) includes several questions relating to walking:

> In the last week, how many times have you walked continuously, for at least ten minutes, for recreation, exercise, or to get to or from places?
> What do you estimate was the total time that you spent walking in this way in the last week?

The international physical activity questionnaire (IPAQ) is a set of standard questionnaires that have been developed to provide a common method for obtaining comparable data across countries. They can be administered by phone or self-administered. The survey and documentation can be downloaded (see: www.ipaq.ki.se). The long survey covers seven days and contains questions asking the frequency (number of days) and duration of cycling and walking.

Conventional travel surveys (self-completion questionnaires or interview surveys) can be used to estimate the amount of time spent walking and cycling. The Victorian Integrated Survey of Travel and Activities (VISTA) is an example of a large-scale transport survey that was recently conducted in Melbourne (DOT 2009). Information on activities undertaken also allows the duration of sport and recreation activities to be estimated.

Geocoding of stops allows direct distances to be estimated for trip stages, but the actual distance traveled depends on the paths traveled, which are not generally provided and must be estimated. It is also difficult to estimate the intensity of physical activity from walking and riding (location of homes) since the exact location of households is not accurately presented due to privacy reasons. The estimation of times and locations is also not generally very accurate and short trips are often under-reported.

Census data can be used to analyze the travel modes used to the journey to work in Australia. This can be conducted at various degrees of spatial aggregation, including local government area and suburban areas (www.censusdata.abs.gov.au). The smallest area of analysis is the census collector district, which is typically around 200 households.

Health surveys can often include questions relating to walking and cycling. For example, the Victorian Lifestyle and Neighborhood Environment Study (VicLANES 2007) investigated the effect of socioeconomic status, neighborhood environments, and individuals' perceptions in relation to food purchasing, physical activity, and alcohol consumption (www.kcwh.unimelb.edu.au/research_old/programs/viclanes). Attitudes and barriers to physical activity were examined for fifty small areas in Melbourne.

3.3.2 Accelerometers

Accelerometers are portable monitors that measure movement in terms of acceleration, which can then be used to record body movements to estimate physical activity levels over an extended period. Acceleration signals are recorded for each spatial axis (anterior–posterior, medial–lateral, and vertical) at high frequency over a range of acceleration levels. For example, a three-axis accelerometer, "alive heart monitor," with a sampling frequency of 75 Hz with an acceleration range of -2.7 to $2.7\,g$ was used to detect walking activity for a cardiac rehabilitation program (Bidargaddi et al. 2007). Triaxial accelerometers record counts for each dimension, while uniaxial accelerometers combine movement counts over all dimensions.

Acceleration is detected as an external force that reflects the energy cost of physical activity. Piezoelectric sensors and seismic mass measure compression upon acceleration and generate a voltage signal proportionate to acceleration. Accelerometers are usually mounted on the waist, close to the center of the mass of a person, to provide information on physical activities of the person.

Recorded counts are aggregated over an epoch (e.g., 1-minute interval). Movement can be classified in intensity categories (e.g., sedentary, light, moderate, and vigorous). Accelerometers measure dynamic activities such as walking, ascending stairs, descending stairs, jogging, and running.

Accelerometers have the ability to measure incidental, intermittent PA such as short walks objectively. They also enable movement to be monitored inside buildings, which is often not recorded and is difficult using other technologies such as GPS. Accelerometers are becoming smaller, cheaper, easier to use and are now feasible for large-scale health surveys. This allows monitoring of physical activities in the field and out of the laboratory.

Accelerometers allow comprehensive profiling of physical activity bouts (Esliger and Tremblay 2007). They also have low subject burden (not having to rely on the memory of individuals), and are unobtrusive, allowing recording over multiple days. Accelerometers record the frequency, duration, and intensity of physical activity, objectively. These data can be used to estimate energy expenditure and physical activity levels.

The Actigraph records activity counts (4 *mg* per second), summed over an epoch (e.g., 1 minute). It can also estimate the number of steps in each epoch. Data from the Actigraph can be downloaded to a personal computer using a cable with a USB connection. The Actigraph is best worn with an elastic belt on the hip.

When a standard 1-minute epoch is used with step mode, it has a memory capacity of 180 days (Actigraph 2008). It has a battery that allows up to fourteen days of data to be recorded without recharging. Activity counts have been linked to intensity categories and METs (Table 3.1).

The Actigraph accelerometer has been used in a number of health studies (U.S. National Health and Nutrition Examination Survey [NHANES]: www.cdc.gov/nchs/nhanes.htm).

Data from accelerometers can be used to perform detailed analysis of physical activity daily activity profiles, as well as the distribution of bouts (Esliger and Tremblay 2007). Analysis can be undertaken using a spreadsheet (Figures 3.4–3.9). If an activity diary is kept, additional analysis of counts can be undertaken to investigate the amount of physical activity gained from transport (Figure 3.8). Data can be used

TABLE 3.1 Activity Counts and METs (1-Minute Epoch)

Activity Count	Intensity Category	METs
<1952	Light	<2.99
[1953, 5724]	Moderate	[3, 5.99]
[5725, 9498]	Hard	[6, 8.99]
>9498	Very hard	>9

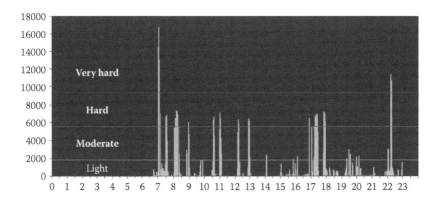

FIGURE 3.4 (See color insert.) Activity count profile.

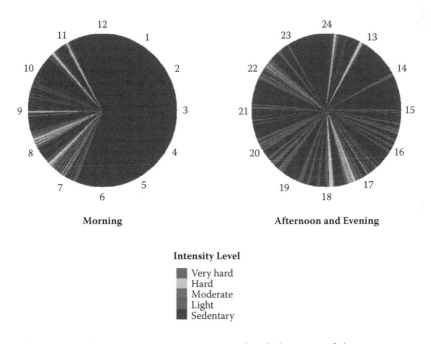

FIGURE 3.5 (See color insert.) Intensity levels by time of day.

to estimate physical activity levels (PALs) as well as checking that health guidelines are being achieved. The frequency and duration of sedentary periods can also be undertaken; this has been linked to health benefits (Healy et al. 2007, 2008).

Accelerometers have some limitations in monitoring physical activity from transport. No information on the context or source of physical

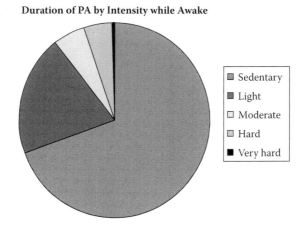

FIGURE 3.6 (See color insert.) Duration of physical activity by intensity levels.

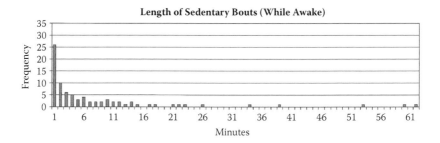

FIGURE 3.7 Distribution of sedentary bouts.

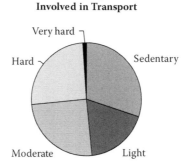

FIGURE 3.8 (See color insert.) Duration of intensity levels for transport.

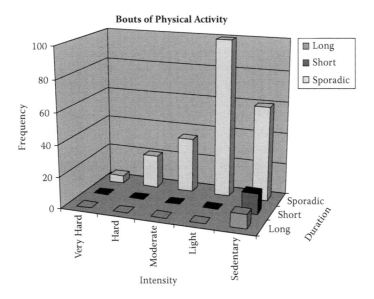

FIGURE 3.9 (See color insert.) Bouts of physical activity by frequency, duration, and intensity.

activity is recorded. To link the count data to transport or other types of physical activity, a diary generally needs to be recorded manually. Although models have been developed for determining the type of physical activity mode (Pober et al. 2006; Bidargaddi et al. 2007; Veltink et al. 1996) from accelerometer data combined with GPS devices (Troped et al. 2008), these procedures are not yet widely used by practitioners.

Accelerometers have been found to underestimate physical activity levels from walking and overestimate physical activity levels from jogging (Shephard 2003). Also, movement of extremities is not detected (e.g., arms) since accelerometers miss upper body movements. They do not measure load bearing, so items carried are not considered.

Subjects may also increase their physical activity levels due to monitoring, although surveys of a week's duration are recommended because they reduce this potential. Compliance in wearing the devices over many days can also be an issue. A major weakness of accelerometers with respect to measuring physical activity from active transport is that they do not record movement associated with bicycle riding.

3.3.3 Global Position Systems

Global position systems allow the time and location of mobile devices to be recorded electronically. In-vehicle navigation systems are a common application of GPS. Data captured from GPS devices can be used to estimate energy expenditure from walking and cycling since the speed and duration of movement can be determined.

However, there are a number of problems related to the application of GPS to estimate the levels of physical activity from transport—such as determining the mode and purpose of the trip as well as some locations that are not recorded due to weak signals. GPS data do not indicate what transport mode a person has used. Techniques for automatically determining the mode of transport are the focus of ongoing research. The purpose of the trip or the activity undertaken at the end of the trip is also not recorded by GPS. This limits the understanding as to why trips were undertaken and the nature of activities at the destinations. Research is currently being undertaken to develop methods for automatically analyzing the data collected from GPS (Stopher, Fitzgerald, and Zhang 2008). Algorithms are required for identifying separate trips as well as determining trip purpose and travel mode used.

In urban areas it is common for the signals from a satellite to be interrupted due to inference from obstructions such as buildings, tunnels, and electricity cables. GPS devices have difficulty receiving signals when they are, inside or near buildings, underground, or near overhead electricity cables. Such situations are common when persons are traveling in urban areas. It is also common for GPS devices to take some time (e.g., several minutes) to determine the location of the device when it is turned on. This can limit the ability of GPS to record short trips accurately (e.g., walking).

3.4 MODELING

3.4.1 Estimating Physical Activity Levels

Daily physical activity levels can be determined by estimating the total energy expenditure, expressed as a multiple of the basal metabolic rate. It is recommended that average PALs should be above 1.6 (AICR 2007).

Estimating energy expenditure for an individual over a daily period is a complex and challenging task. There are several methods of calculating physical activity levels. A simple method has developed for determining this that is based on personal attributes such as age, height, weight, and gender as well the duration and metabolic rate of activities undertaken (Gerrior, Juan, and Basiotis 2006):

$$BEE_i = 293 - 3.8a_i + 456.4h_i + 10.12w_i \quad \text{if } g_i = male \quad (3.1)$$

$$BEE_i = 247 - 2.67a_i + 401.5h_i + 8.6w_i \quad \text{if } g_i = female \quad (3.2)$$

$$\Delta PAL_{ij} = \frac{[(MET_j - 1) \times (1.15 / 0.9) \times d_{ij}] / 1440}{BEE_i / (0.0175 \times 1440 \times w_i)} \quad (3.3)$$

$$PAL_i = 1.1 + \sum_j \Delta PAL_{ij} \quad (3.4)$$

where
BEE_i = basal energy expenditure for person i
a_i = age of person i (years)
h_i = height of person i (meters)
w_i = weight of person i (kilograms)
g_i = gender of person i (male or female)
MET_j = metabolic equivalence of activity j
d_{ij} = duration of activity j for person i (minutes)
ΔPAL_{ij} = energy expenditure of activity j for person i

A number of categories have been determined by the Food and Nutrition Board (2005) for PALs (Table 3.2).

To estimate PALs it is necessary to determine the MET for each of the activities and travel modes. METs are defined as the metabolic work rate expressed as a multiple of the standard resting rate.

Ainsworth et al. (2000) presented 605 METs for twenty-one activity categories at a variety of intensity levels. The reported METs range from 1 for riding in a car or truck to 18 for running at 17.4 km/h. For walking, a total of 43 METs were presented at various speeds, purposes, and loads carried. A number of examples relevant to walking for transport are presented in Table 3.3.

Cycling can be for leisure, work, or sport at various levels of intensity and speeds. Several MET examples for common forms of cycling for transport are presented in Table 3.4.

Procedures have been developed for combining data from accelerometers and GPS to estimate the amount of physical activity over a weekly period (Thompson and Kayak 2011). This allows an accurate assessment as to whether the health guidelines are being achieved and the contribution of transport to daily energy expenditure.

TABLE 3.2 PAL Categories and Ranges

PAL Category	PAL Range
Sedentary	1–1.39
Low active	1.4–1.59
Active	1.6–1.89
Very active	1.9–2.49

TABLE 3.3　Walking MET Examples

Activity	METs
Speed less than 3.2 km/h, level ground, strolling very slowly	2
From house to car or bus, from car or bus to go places, from car or bus to and from the worksite	2.5
Speed at 5 km/h, level, moderate pace and firm surface	3.3
Carrying 6.8 kg bag on level ground	3.5
Speed of 8 km/h	8

Source: Ainsworth, B. E. et al. (2000). *Medicine & Science in Sports & Exercise* 32: S498–S516.

TABLE 3.4　Cycling MET Examples

Activity	METs
For leisure or work < 16 km/h	4
General	8
Fast, vigorous effort (22–25 km/h)	10

Source: Ainsworth, B. E. et al. (2000). *Medicine & Science in Sports & Exercise* 32: S498–S516.

3.4.2 Energy Balance Modeling

It possible to predict the weight of individuals based on assumed levels of energy intake and expenditure (Payne and Dugdale 1977; Westerterp et al. 1995). Scenarios of increased physical activity over prolonged periods can be investigated in terms of weight loss.

3.5 EVALUATION

3.5.1 Benefit-Cost Analysis

An important element of any engineering system is evaluating alternatives. The economic costs and benefits of walking and cycling tracks were estimated for several Norwegian cities (Saelensminde 2004). A number of benefits were identified and quantified, including reduced costs associated with:

- Insecurity for new and current pedestrians and cyclists
- Transporting school children
- Diseases and ailments

- External costs of motorized road transport
- Parking costs for employees

The benefits estimated for cycle track facilities were at least four times the costs. A substantial proportion of the benefits (over 50%) were associated with the reduced costs related to severe diseases and ailments. Only four types of diseases or ailments were included: cancers, high blood pressure, diabetes type 2, and musculoskeletal ailments.

Cycling leads to a reduction in car use, thus reducing road safety costs, externality, and congestion costs (Bauman et al. 2008). Health and fitness benefits of A$0.376/cycle-kilometer (mortality plus morbidity) were estimated recently.

3.5.2 Multicriteria Methods

A multicriteria framework for assessing proposals for new bicycle facilities has been developed (ARRB 2005). This framework has a number of components (Figure 3.10).

The required information of the framework consists of a route diagram, cost, and maintenance arrangements. The route diagram highlights the connectivity of the facility and the proximity to demand generators. Costs include whole-of-life costs including capital and maintenance costs. The project priorities matrix consists of a number of metrics that

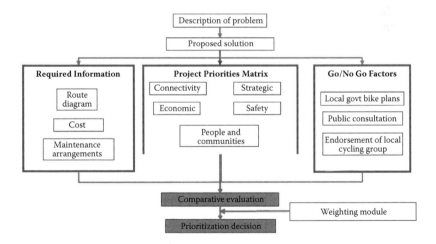

FIGURE 3.10 Bicycle infrastructure prioritization framework. (ARRB, 2005, Prioritization of bicycle infrastructure proposals, WC5227, December 2005, ARRB Vermont South.)

assess the objectives and subobjectives. The go/no go factors cover "must haves" that are considered essential for any cycling facility proposal. Checklists for "bikeability" for safety and level of service have been developed. Strategic objectives largely relate to the completion of state networks. Connectivity involves proximity to employment zones, schools, recreational/tourism facilities, and public transport. Economic objectives include mode shift (reduction in private car travel) and impact on motor vehicles (speeds and operating costs). Safety objectives include cycling safety and pedestrian safety issues.

A multicriteria assessment module has also been defined for combining weightings and performance for the strategic, connectivity, economic, safety, and "people and communities" objectives (ARRB 2005).

With commuter bicycle use in the Melbourne statistical division (MSD) dramatically increasing, and Bauman et al. (2008) estimating that commuter cyclists currently save the economy $72.1 million per year in reduced health costs, there is strong justification for investment in research to measure the health benefits from increased use of both the main modes of active transport. The overall direct gross cost of physical inactivity on the Australian health budget in 2006/2007 was $1.49 billion according to a recent study (Econtech 2007).

REFERENCES

Actigraph (2008). ActiGraph GT1M Monitor/ActiTrainer and ActiLife Lifestyle Monitor software user manual, ActiGraph, LLC Pensacola, FL (www.theactigraph.com).

AICR (2007). *Food, nutrition, physical activity, and the prevention of cancer: A global perspective.* World Cancer Research Fund/American Institute for Cancer Research, Washington, DC: AICR.

Ainsworth, B. E., Haskell, W. L., Whitt, M. C., Irwin, M. L., Swartz, A. M., Strath, S. J., O'Brien, W. L., Bassett, D. R., Jr., Schmitz, K. H., Emplaincourt, P. O., et al. (2000). Compendium of physical activities: An update of activity codes and MET intensities. *Medicine & Science in Sports & Exercise* 32: S498–S516.

ARRB (2005). Prioritization of bicycle infrastructure proposals, WC5227, December 2005, ARRB Vermont South.

Australian Institute of Health and Welfare (2003). *The active Australia survey, a guide and manual for implementation, analysis and reporting.* Canberra.

AUSTROADS (1995). Guide to traffic engineering practice part 13—Pedestrians. Publication no. AP-11.13/95, Sydney.

——— (1999). Guide to traffic engineering practice part 14—Bicycles. Publication no. AP-11.14/99, Sydney.

Bauman, A., Rissel, C., Garrard, J., Ker, I., Speidel, R., and Fishman, E. (2008). Cycling: Getting Australia moving: Barriers, facilitators and interventions to get more Australians physically active through cycling. Cycling Promotion Fund, Melbourne.

Berrigan, D., Troiano, R. P., McNeel, T., DiSogra, C., and Ballard-Barbash, R. (2006). Active transportation increases adherence to activity recommendations. *American Journal of Preventive Medicine* 31 (3): 210–216.

Bidargaddi, N., Sarela, A., Klingbeil, L., and Karunanithi, M. (2007). Detecting walking activity in cardiac rehabilitation by using accelerometer. *Intelligent Sensors, Sensor Networks and Information 2007*, ISSNIP 2007, IEEE, December 3–6, 2007, Melbourne, pp. 555–560.

Blaire, S., and Brodney, S. (1999). Effects of physical inactivity and obesity on morbidity and mortality: Current evidence and research issues. *Medical Science and Sports Exercise* 31: S686–S692.

Booth, M. L., Wake, M., Armstrong, T., Chey, T., Hesketh, K., and Mathur, S. (2001). The epidemiology of overweight and obesity among Australian children and adolescents, 1995–97. *Australia New Zealand Journal Public Health* 25: 162–169.

Bouchard, C., and Shephard, R. (1994). Physical activity fitness, and health: The model and key concepts. In *Physical activity fitness and health*, ed. C. Bouchard, R. Shephard, and T. Stephens, 77–88. International Proceedings and Consensus Statement, Champaign, IL: Human Kinetics.

Bull, F. C., Bauman, A., Brown, W. J., and Bellew, B. (2004). Getting Australia Active II: An update of evidence on physical activity for health. National Public Health Partnership (NPHP), Melbourne.

Davison, K. K., and Campbell, K. (2005). Opportunities to prevent obesity in children within families: An ecological approach. In *Obesity prevention and public health*, ed. D. Crawford and R. W. Jeffery. Oxford, England: Oxford University Press.

Department of Health and Children, Health Service Executive (2009). The national guidelines for physical activity in Ireland, Get Ireland Active. http://www.getirelandactive.ie/pdfs/GIA_GUIDE.pdf

DHS (2005). Victorian burden of disease study: Mortality and morbidity in 2001. Department of Human Services, Melbourne.

DOT (2009). Victorian integrated survey of transport and activity (VISTA). Department of Transport, Victorian government, Melbourne. http://www.transport.vic.gov.au/Doi/Internet/planningprojects.nsf/All Docs/7E437B50F6AACA38CA25728000146838?OpenDocument

Econtech (2007). Economic modeling of the net costs associated with non-participation in sport and physical activity. Prepared by Econtech Pty. Ltd. for Medibank Private.

Esliger, D. W., and Tremblay, M. S. (2007). Physical activity and inactivity profiling: The next generation. *Applied Physiology, Nutrition, and Metabolism* 32: S195–S207.

Food and Nutrition Board (2005). *Dietary reference intakes for energy, carbohydrate, fiber, fat, fatty acids, cholesterol, protein, and amino acids (macronutrients)*. Washington, DC: National Academy Press.

Frank, L. D., Schmid, T. L., Sallis, J. F., Chapman, J., and Saelens, B. E. (2006). Linking objectively measured physical activity with objectively measured urban form; findings from SMARTRAQ. *American Journal of Preventive Medicine* 28: 117–125.

Gerrior, S., Juan, W., and Basiotis, P. (2006). An easy approach to calculating estimated energy requirements. *Prevention of Chronic Disease* 3 (4). http://www.pubmedcentral.nih.gov/articlerender.fcgi?artid = 1784117 (retrieved May 30, 2008).

Handy, S. (2004). Critical assessment of the literature on the relationship among transportation, land use and physical activity. Prepared for the Transportation Research Board and the Institute of Medicine Committee on Physical Activity, Health, Transportation, and Land Use, Washington, DC (http://trb.org/downloads/sr282papers/sr282 Handy.pdf).

Haskell, W., Lee, I., Pate, R., Powell, K., Blair, S., Franklin, B., et al. (2007). Physical activity and public health: Updated recommendation for adults from the American College of Sports Medicine and the American Heart Association. *Medical Science and Sports Exercise* 39: 1423–1434.

Healthy Living Unit. (1993). What is active transportation? The Public Health Agency of Canada. http://www.phac-aspc.gc.ca/pau-uap/fitness/active_trans.htm (retrieved May 30, 2008).

Healy, G. N., Dunstan, D. W., Salmon, J., Cerin, E., Shaw, J. E., Zimmet, P. Z., and Owen, N. (1997). Objectively measured light-intensity physical activity is independently associated with 2-h plasma glucose. *Diabetes Care* 30 (6): 1384–1389.

——— (1998). Breaks in sedentary time—Beneficial associations with metabolic risk. *Diabetes Care* 31 (4): 661–666.

Hu, W., Thompson, R. G., and Zaman, R. (2008). An active transport journey planner methodology. *30th Conference of Australian Institutes of Transport Research (CAITR)*, December 2008, University of Western Australia, Perth.

Li, M. W., and Rissel C. (2008). Inverse associations between cycling to work, public transport, and overweight and obesity: Findings from a population based study in Australia. *Preventive Medicine* 46 (1): 29–32.

Murray, C. (1994). Quantifying the burden of disease: The technical basis for disability-adjusted life years. *Bulletin of the World Health Organization* 72: 429–445.

Murray, C. J., and Acharya, A. K. (1997). Understanding DALYs (disability-adjusted life years). *Journal of Health Economics* 16: 703–730.

NPHP (National Public Health Partnership) (2001). Promoting active transport: An intervention portfolio to increase physical activity as a means of transport. National Public Health Partnership.

Newman, P., and Kenworthy, J. (1999). *Sustainability and cities: Overcoming automobile dependence.* Washington, DC: Island Press.

Owen, N., Cerin, E., Leslie, E., duToit, L., Coffee, N., Frank, L. D., Bauman, A. E., Hugo, G., Saelens, B. E., and Sallis, J. F. (2007). Neighborhood walkability and the walking behavior of Australian adults. *American Journal of Preventive Medicine* 33 (5): 387–395.

Payne, P. R., and Dugdale, A. E. (1977). A model for the prediction of energy balance and body weight. *Annals of Human Biology* 4 (6): 525–535.

Pober, D. M., Staudenmayer, J., Raphael, C., and Freedson, P. S. (2006). Development of novel techniques to classify physical activity mode using accelerometers. *Medicine and Science in Sports and Exercise* 38: 1626–1634.

Prochaska, J. O., and DiClemente, C. C. (1983). Stages and processes of self-change in smoking: Towards an integrative model of change. *Journal of Consulting and Clinical Psychology* 51: 390–395.

Prochaska, J. O., Norcross, J. C., and Diclemente, C. (1994). *Changing for good.* New York: Quill.

Sælensminde, K. (2004). Cost-benefit analyses of walking and cycling track networks taking into account insecurity, health effects and external costs of motorized traffic. *Transportation Research Part A* 38: 593–606.

Sallis, J., Bauman, A., and Pratt, M. (1998). Environmental and policy interventions to promote physical activity. *American Journal of Preventative Medicine* 15: 379–397.

Salmon, J. A., and King, A. C. (2005). Population approaches to increasing physical activity among children and adults. In *Obesity prevention and public health,* ed. D. Crawford and R. W. Jeffery, 129–152. Oxford, England: Oxford University Press.

Shephard, R. J. (2003). Limits to the measurement of habitual physical activity by questionnaires (review). *British Journal of Sports Medicine* 37 (3): 197–210.

——— (2008). Is active commuting the answer to population health? *Sports Medicine* 38 (9): 751–758.

Stopher, P., Fitzgerald, C., and Zhang, J. (2008). Search for a global positioning system device to measure person travel. *Transportation Research Part C* 16: 350–369.

Thompson, R. G., and Kayak, H. (2011). Estimating personal physical activity from transport. *Proceedings Australasian Transport Research Forum 2011,* September 28–30, Adelaide. http://www.patrec.org/atrf.aspx

Troped, P., Oliveira, M. S., Matthews, C. E., Cromley, E. K., Melly S. J., and Craig, B. A. (2008). Prediction of activity mode with global positioning system and accelerometer data. *Medicine and Science in Sports and Exercise* 40: 972–978.

U.S. Department of Health and Human Services (2008). *2008 Physical activity guidelines for Americans.* www.health.gov/paguidelines

Veltink, P. H., Bussmann, H. B. J., de Vries, W., Martens, W. L. J., and Van Lummel, R. C. (1996). Detection of static and dynamic activities using uniaxial accelerometers. *IEEE Transactions on Rehabilitation Engineering* 4 (4): 375–385.

VicLANES (2007). Victorian lifestyle and neighborhood environment study. University of Melbourne. http://www.kcwh.unimelb.edu.au/research/major_research_programs/VicLANES

Westerterp, K. R., Donkers, J. H. H. L. M., Fredrix, E. W. H. M., and Boekhoudt, P. (1995). Energy intake, physical activity and body weight: A simulation model. *British Journal of Nutrition* 73: 337–347.

WHO (2006). *Guidelines for the prevention, management and care of diabetes mellitus,* ed. O. M. N. Khatib. EMRO technical publications series 32, World Health Organization, Regional Office for the Eastern Mediterranean.

CHAPTER 4

Hazardous Material Transportation

Wai Yuen Szeto and Rojee Pradhananga

CONTENTS

4.1 Hazardous Materials: Definition and Classification 77
4.2 The Cause, Significance, and Mode of Hazmat Shipments 79
4.3 Hazmat Shipment Accidents: Causes, Consequences, and Probability of Occurrence 79
4.4 Multiple Parties and Multiple Objective Hazmat Shipment 84
4.5 Risk Management 85
    4.5.1 Regulations 85
    4.5.2 Compliance 86
    4.5.3 Exemptions 86
    4.5.4 Mitigation 86
    4.5.5 Risk Management Self-Evaluation Framework 87
4.6 Risk Assessment 87
4.7 Routing 90
    4.7.1 Optimization Techniques for Hazmat Routing Problems 92
        4.7.1.1 Shortest Path Models 92
        4.7.1.2 Vehicle Routing Problem Models 94
4.8 Combined Hazmat Facility Location and Routing Problem 95
4.9 Concluding Remarks 96
References 96

4.1 HAZARDOUS MATERIALS: DEFINITION AND CLASSIFICATION

A hazardous material (or hazmat for short) is defined as any substance or material capable of causing harm to people, property, or the environment (U.S. Department of Transportation [U.S. DOT]). Some daily life examples include gasoline and diesel. Indeed, there are many hazmats in reality. They are classified into nine categories according to their physical, chemical, and nuclear properties (Keller and Associates 2001). Table 4.1 depicts the nine categories in detail. As we can see from this table, hazmats can be explosive, flammable, oxidizing, toxic, radioactive, or corrosive.

TABLE 4.1 Classification of Hazmats

Class	Hazard
Class 1	**Explosives**
Division 1.1	Explosives with a mass explosion hazard
Division 1.2	Explosives with a projection hazard
Division 1.3	Explosives with predominately a fire hazard
Division 1.4	Explosives with no significant blast hazard
Division 1.5	Very sensitive explosives; blasting agents
Division 1.6	Extremely insensitive detonating devices
Class 2	**Gases**
Division 2.1	Flammable gases
Division 2.2	Nonflammable, nontoxic compressed gases
Division 2.3	Gases toxic by inhalation
Class 3	**Flammable (and combustible) liquids**
Class 4	**Flammable solids**
Division 4.1	Flammable solids
Division 4.2	Spontaneously combustible materials
Division 4.3	Dangerous when wet materials
Class 5	**Oxidizers and organic peroxides**
Division 5.1	Oxidizers
Division 5.2	Organic peroxides
Class 6	**Toxic materials and infectious substances**
Division 6.1	Poisonous materials
Division 6.2	Infectious substances
Class 7	**Radioactive materials**
Class 8	**Corrosive materials**
Class 9	**Miscellaneous dangerous goods**

4.2 THE CAUSE, SIGNIFICANCE, AND MODE OF HAZMAT SHIPMENTS

Shipment of hazmats happens every day and everywhere including the United States, Canada, and the Asian megacities. The locations of the demand for hazmats are normally not next to the locations of the supply and the locations for disposal of hazmat wastes. Transportation is required to ship the hazmats from the supply locations to the demand locations and from the demand location to the disposal locations.

The number and weight of hazmat shipments in the United States has increased over time. The Office of Hazardous Materials Safety (OHMS) of the U.S. DOT estimated that there were 800,000 domestic shipments of hazmats, totaling approximately nine million tons, in the United States each day in 1998, which is equivalent to 3.1 billion tons shipped in that year (U.S. DOT 1998a). In 2002, in the United States alone, the annual tonnage of hazmat shipped was increased to 2.19 billion in total, as shown in Table 4.2; this represents 18.9% of total goods shipped (U.S. DOT 2004a). In 2009, almost one million daily shipments were estimated to be transported in the United States (Wikipedia 2009).

Because of constraints of geographical locations and cost, hazmats are not restricted to transport by only one mode, such as truck, rail, air, water, or pipeline, but may be transported by more than one mode. Table 4.2 shows the hazardous material shipment characteristics by mode of transportation for the United States in 2002 (U.S. DOT 2004a). From this table, we can see that trucks carried more than 60% of hazmats by value and 50% by tonnage in 2002. However, average miles per shipments by trucks are lower than those for rail and air, possibly because trucks are normally used for last-mile shipment with short travel distances.

The number of hazmat shipments depends on the size of a country and its level of industrialization (Erkut, Tjandra, and Verter 2007). Nearly 80,000 shipments of dangerous goods were transported daily in Canada in 2004 (Transport Canada 2004)—10 times lower than such shipments in the United States in 1998.

4.3 HAZMAT SHIPMENT ACCIDENTS: CAUSES, CONSEQUENCES, AND PROBABILITY OF OCCURRENCE

It is not surprising that accidents related to hazmats do occur, given that massive shipments are transported daily. For example, train derailments in Alberton, Montana, in 1996, and in Graniteville, South Carolina, in

TABLE 4.2 Hazardous Material Shipment Characteristics by Mode of Transportation for the United States in 2002

Mode of Transportation	Value		Tons		Ton-miles		Average Miles per Shipment
	Million Dollars	Percent	Thousands	Percent	Millions	Percent	
For-hire truck	189,803	28.8	449,503	20.5	65,112	19.9	285
Private truck	226,660	34.3	702,186	32	44,087	13.5	38
Rail	31,339	4.7	109,369	5	72,087	22.1	695
Water	46,856	7.1	228,197	10.4	70,649	21.6	S
Air (includes truck and air)	1,643	0.2	64	—	85	—	2,080
Pipeline (excludes crude petroleum)	145,021	22	661,390	30.2	S	S	S
Multiple modes	9,631	1.5	18,745	0.9	12,488	3.8	849
Other and unknown modes	6,061	0.9	14,241	0.6	2,342	0.7	57

Source: US DOT. (2004a). United States: 2002 Hazardous materials. Bureau of Transportation Statistics and U.S. Census Bureau, Washington, DC. Available at http://www.bts.gov/publications/commodity_flow_survey/2002/hazardous_materials/pdf/entire.pdf

2005, resulted in release of chlorine into the atmosphere. The accidents that result in releasing hazmats are called incidents.

The incidents can occur anywhere along the trips, including the origins and destinations, due to loading and unloading hazmats respectively, but these incidents only correspond to a small proportion of traffic accidents. The annual number of U.S. transportation accidents was about six million in 2008 (U.S. DOT 2008b), in contrast to the approximately 20,340 hazmat transportation incidents in the same year (U.S. DOT 2009).

Among various categories of hazmats, flammable–combustible liquid accounted for the majority (more than 50%) of hazmat incidents in the United States in 2008 (see Figure 4.1). Corrosive materials are in the second place (25%), combustible liquid is in the third place (9%), and nonflammable gas is in fourth place (5%). All these facts imply that flammable combustible liquid and corrosive materials account for most of the incidents.

The highway mode accounts for most of the incidents in North America. During the period between 1999 and 2008 in the United States, most of the incidents were on highways (see Figure 4.2). In particular, about 86% of the total of 17,335 U.S. incidents in 2008 were highway related (U.S. DOT 2008b). A similar figure (89% of the total of 127 incidents) has been reported by Transport Canada (2008). This high

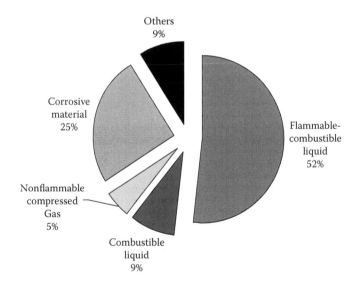

FIGURE 4.1 (See color insert.) Accident/incident by HAZMAT class in 2008. (US DOT, 2008a, Pipeline and Hazardous Materials Safety Administration, U.S. Department of Transportation, Washington, DC. http://www.phmsa.dot.gov/staticfiles/PHMSA/DownloadableFiles/Files/2008comdrank.pdf.)

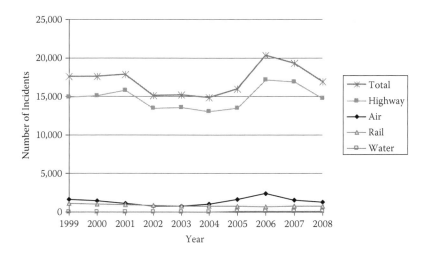

FIGURE 4.2 The number of incidents by mode of transport during 1999–2008. (US DOT, 2009, Ten year hazardous materials incident data. U.S. Pipeline and Hazardous Materials Safety Administration. U.S. Department of Transportation. Available at http://www.phmsa.dot.gov/staticfiles/PHMSA/DownloadableFiles/Files/tenyr.pdf.)

percentage of highway incidents was more or less the same throughout the 2003–2007 period (Transport Canada 2008).

Incidents can lead to injuries and fatalities. Figures 4.3 and 4.4 give the numbers of fatalities and injuries by mode during the period of 1999–2008 in the United States. The highway mode again accounts for most fatalities in the reported period. During the period between 2003 and 2007 in Canada, on average, seven people were killed and forty-two people were injured annually (Transport Canada 2008).

The economic loss from incidents can be high and the incidents can also lead to traffic disruption. According to the U.S. DOT (2009), the total property damage cost was about $44 million in 2008. Table 4.3 shows the costs of hazmat and nonhazmat motor carrier accident/incident events (Federal Motor Carrier Safety Administration 2001). This table shows that the costs of incidents associated with fire and explosion are much higher than those for incidents associated with spill/release and the cost of nonhazmat accidents. This table also shows that the traffic delay induced by the incidents with fire and explosion is also longer than that from incidents with spill/release and nonhazmat accidents.

The causes of incidents include human errors, packing failures, the inadequate maintenance of transport vehicles, etc. Among these causes, human errors seems to be the single greatest factor in all hazardous

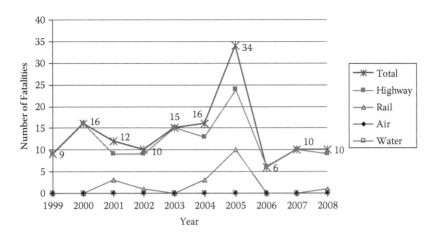

FIGURE 4.3 The number of fatalities due to incidents during 1999–2008 (U.S. DOT, 2009, Ten year hazardous materials incident data. U.S. Pipeline and Hazardous Materials Safety Administration. U.S. Department of Transportation. Available at http://www.phmsa.dot.gov/staticfiles/PHMSA/DownloadableFiles/Files/tenyr.pdf.)

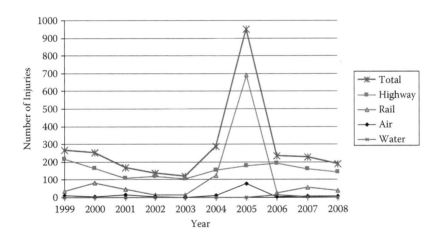

FIGURE 4.4 The number of injuries due to incidents during 1999–2008 (US DOT, 2009, Ten year hazardous materials incident data. U.S. Pipeline and Hazardous Materials Safety Administration. U.S. Department of Transportation. Available at http://www.phmsa.dot.gov/staticfiles/PHMSA/DownloadableFiles/Files/tenyr.pdf.)

TABLE 4.3 Comparative Costs of Hazmat and Nonhazmat Motor Carrier Accident/Incident Events

Type of Accident/Incident Event	Average Cost (in US$)	Average Traffic Delay (in hours)
Nonhazmat events	340,000	2
All hazmat events	414,000	—
Hazmat events with spill/release	536,000	5
Hazmat events with fire	1,200,000	8
Hazmat events with explosion	2,100,000	12

Source: Federal Motor Carrier Safety Administration (2001). Comparative risks of hazardous materials and non-hazardous materials truck shipment accidents/incidents. Washington, DC.

materials incidents in 2003 (U.S. DOT 2003c) and remain one of the greatest factors nowadays.

Given that the daily hazmat shipment is about one million and the annual number of incidents is less than 20,000, the probability of having an incident is less than $20,000/1,000,000/365 = 7.5 \times 10^{-5}$. Because of this low probability and highly undesirable consequence of incidents, the term "low-probability, high-consequence events" is normally used to describe incidents.

4.4 MULTIPLE PARTIES AND MULTIPLE OBJECTIVE HAZMAT SHIPMENT

According to Erkut et al. (2007), hazmat transportation involves multiple parties, such as shippers, carriers, packaging manufacturers, freight forwarders, consignees, insurers, governments, and emergency responders. Each has a different role in safely moving hazmats from their origins to their destinations; there are often multiple handoffs of hazmats from one party to another during transport.

Various parties usually have different objectives. Carriers aim at selecting lowest cost routes for shipment. These routes may pass through highly populated areas. If an incident occurs on these routes, the surrounding population can be highly affected. Therefore, the government aims at designating allowable routes that reduce the impact of incidents. This objective is different from that of individual carriers.

The government is also required to deal with the public opposition that can be due to public assessment of risks, equity concerns, and the fears of terrorist attacks on hazmat vehicles. Public assessment of a risk depends on both its magnitude and subjective perceptions. The magnitude of risk can be small but due to the amplification of the risk by mass media, the subjective perception of risk magnitude can be large. Therefore, a small magnitude of risk can result in a strong public opposition of allowing certain routes for transporting hazmats.

The public opposition can also be due to equity concerns. The people benefiting from hazmat shipments are normally living close to the supply or demand of hazmat locations, but the risks are imposed on the people living along the routes connecting supply and demand locations. The people being exposed to the risks are normally not those benefiting from the shipment.

Public opposition to certain highly populated routes can be due to the worry of terrorist attacks on hazmat vehicles. If the hazmat vehicles on the highly populated routes are attacked, the number of fatalities must be huge. Consequently, the routes must be designed to be far from highly populated areas.

4.5 RISK MANAGEMENT

Because of the high consequence of a hazmat incident, the government has to manage the risk carefully. This is not an easy task as multiple parties involved in hazmat transport have different objectives and there can be strong public opposition for hazmat shipment design. The government is required to consider the objective of each party before doing risk management.

According to the U.S. DOT (1998b),

> Risk management is the systematic application of policies, practices and resources to the assessment and control of risk affecting human health and safety and the environment. Hazard, risk and benefit/cost analysis are used to support development of risk reduction options, program objectives, and prioritization of issues and resources. Performance measures are monitored to support performance evaluation.

Risk management has four key elements: regulations, compliance, exemptions, and mitigation (U.S. DOT 1998b). The following four sections provide a summary of the U.S. DOT's (1998b) stance on these elements.

4.5.1 Regulations

The objectives of introducing regulations are to

- Provide safety for workers and the public
- Protect property and the environment
- Minimize cost to the public industry and government
- Minimize economic and social disruption

Currently, the 15th revised edition of *UN Recommendations on the Transport of Dangerous Goods* (United Nations 2007) is considered a model regulation. It covers many aspects, including the following:

- The classification of dangerous goods, their listing, use, and construction
- Testing and approval of packagings and portable tanks
- Consignment procedures such as marking, labeling, placarding, and documentation

The document will provide structural harmony of transport regulations and easier adoption within national legislation of countries throughout the world (Hancock 2001).

4.5.2 Compliance

Ensuring that industry and the public are aware of risks and regulations to control risks can be done by

- Outreach
- Training
- Information dissemination
- Enforcement

Enforcement places greater emphasis on high-risk materials and activities (e.g., toxic and flammable gases, toxic inhalation hazard liquids, explosives, and air transportation). When noncompliance is found, risk and benefit/cost assessments are used to select an appropriate course of action to protect public safety (e.g., safety notices, recalls, down-rating, and use restriction).

4.5.3 Exemptions

Exemptions are alternative risk-based regulations to allow rapid implementation of new technologies and more efficient transportation operations. They provide relief from regulations when exceptions to the rule are allowed for special circumstances. They are granted when an alternative is demonstrated to present a level of safety (risk) equivalent to that provided by the regulations. They are granted on a case-by-case basis.

4.5.4 Mitigation

In terms of mitigation, aids exist in the United States to reduce the impact of incidents on affected people and injuries. For example, the *Handbook* of *Chemical Hazards Analysis Procedures* and the *Automated Resource for Chemical Hazard Incident Evaluation* (ARCHIE) are risk analysis tools to support emergency preparedness and response planners. The *North American Emergency Response Guidebook* provides guidance for first responders. Moreover, planning and training grants are funds to state and local emergency preparedness and emergency response organizations for planning and training.

4.5.5 Risk Management Self-Evaluation Framework

Figure 4.5 provides a self-evaluating risk management framework proposed by ICF Consulting (2000). This framework aims at helping involved parties systematically think about and manage, in a cost-effective manner, the risks associated with transportation of hazardous materials. The Pipeline and Hazardous Materials Safety Administration's (PHMSA) OHMS encourages those involved in hazardous materials transportation to use this framework as a tool, as appropriate, in their quest to improve safety.

4.6 RISK ASSESSMENT

In risk management (for example, in Figure 4.5), risk assessment is involved. This section summarizes the key points of risk assessment

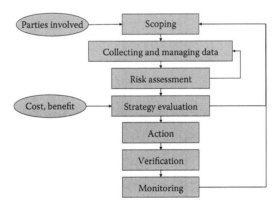

FIGURE 4.5 Risk management self-evaluation framework recommended by ICF Consulting (2000) for all parties involved in hazmat shipment for self-evaluation of their own risks. (ICF Consulting, 2000. Risk management framework for hazardous materials transportation. Submitted to Research and Special Programs Administration, US Department of Transportation.)

described by Erkut et al. (2007). Risk assessment can be classified into two types: qualitative and quantitative. Qualitative risk assessment aims to

- Identify possible accident scenarios
- Estimate undesirable consequences (injuries, fatalities, property damages, etc.)
- Identify the mostly likely events and most severe consequences for further analysis and routing planning

Qualitative risk assessment is applied when there are inaccurate or insufficient data to do quantitative assessment.

Quantitative risk assessment is a numerical assessment of risks involved and consists of the following key steps:

- Hazard and exposed receptor identification
- Frequency analysis
- Consequence modeling
- Risk calculation
- Risk cost determination

The identification of a hazard refers to identifying the potential sources of release of contaminants into the environment, the types and quantities of compounds that are emitted or released, and the potential health and safety effects associated with each substance.

Frequency analysis involves determining the following probabilities for risk calculation:

- The probability $P(A)$ of the occurrence of an accident A (e.g., a crash) related to hazmat transporters
- The probability $P(R|A)$ of the occurrence of a hazmat release R, given that the accident A occurs
- The probability $P(I|R, A)$ of an incident I (e.g., a fire) to be observed, given that a release accident occurs or given that the accident A results in the hazmat release R
- The probability $P(C|A, R, I)$ of the occurrence of an undesirable consequence C, given that the release R due to the accident A results in the incident I to be observed

Note that most of these are conditional probabilities.

In consequence modeling, a consequence is modeled as a function of the impact area. Figure 4.6 shows four shapes of the impact area used in the literature. The basic assumption in this modeling is that each individual within the impact area will be impacted equally and no one outside this area will be impacted.

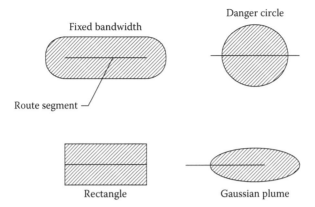

FIGURE 4.6 Common shapes of the impact area used in the literature. (Erkut, E. et al., 2007. In *Handbooks in Operations Research and Management Science*, vol. 14, ed. C. Barnhart and G. Laporte, chap. 9, 539–621. London: Elsevier.)

To calculate the risk associated with an accident, we first determine the probability of the occurrence of an undesirable consequence resulting from a hazmat release on a road segment by

$$p(A, R, I, C) = p(C|A, R, I)p(I|A, R)p(R|A) p(A) \quad (4.1)$$

Note that this equation is valid for any hazmat type m, road segment k, incident type I (e.g., a spill, a fire, or an explosion), accident type A (e.g., crash or derailment), and consequence type C.

Then, we determine the frequency of the occurrence of the release of hazmat m on road segment k by $s_{km}p_k (A, R_m, I, C)$, where s_{km} denotes the number of shipments of hazmat m on road segment k per year.

The risk on road segment k of hazmat m, R_{km}, can then be obtained by

$$R_{km} = s_{km}p_k (A, R_m, I, C)POP \quad (4.2)$$

where POP is the number of people affected by the incident.

To estimate the cost of a hazmat release incident, various consequences must be considered. The consequences can be categorized into

- Injuries and fatalities (or often referred to as population exposure)
- Cleanup costs
- Property damage
- Evacuation

- Product loss
- Traffic incident delay
- Environmental damage

All consequences must be converted to monetary values to permit comparison and calculation of the total risk cost. For example, POP multiplied by the cost of each injury of the incident gives the risk cost for the injuries of the incident.

Let $CONS_c$ be the cost of consequence type c. The total risk cost on road segment k can then be calculated by

$$R_k = \sum_a \sum_m \sum_i \sum_c s_{km} p_k(A_a, R_m, I_i, C_c) CONS_c \qquad (4.3)$$

As reflected from the preceding equation, the total risk cost is obtained by summing up the individual risk cost over all accident types a, hazmat types m, incident levels i, and consequence types c.

4.7 ROUTING

Routing is a problem for hazmat carriers as well as the government. The government requires determining which routes and which lanes the hazmat carriers can use during a particular time period based on the following factors:

- Types and quantities of hazmats
- Results of consultations
- Route continuity
- Effects on commence
- Delays in transportation
- Accidents' history
- Emergency response capability
- Terrain considerations
- Alternative route availability
- Congestion

The carriers determine routes to minimize their costs or to maximize their profit, subject to regulation constraints.

In the literature, there are three classes of problems related to routing (Erkut et al. 2007):

- The local route planning problem
- The local route and schedule planning problem
- The global route planning problem

The local route planning problem is to choose the route(s) between a given origin–destination pair for a given hazmat, a given transportation mode, and a given vehicle type. This problem focuses on a single commodity and a single origin–destination routing decision for each shipment order. This problem also belongs to a carrier's problem since, from the carrier's perspective, shipment contracts can be considered independently and a routing decision needs to be made for each shipment. In the literature, this routing problem can be classified based on three dimensions:

- Transportation modes
- Deterministic/stochastic
- Single/multiple objectives

The local route and schedule planning problem captures the dynamic nature of the routing problem. This problem allows hazmat vehicles departing at various time periods and stopping at intermediate locations to avoid high-risk locations over certain time periods. This problem can be modeled as a path selection problem in a stochastic time-varying network and can be classified into three categories:

- A priori optimization
- Adaptive route selection
- Adaptive route selection with real-time updates

Table 4.4 depicts the differences between various local route and schedule planning problems. Like the local routing problem, local routing and scheduling decisions are often made without taking into consideration the overall impact on society and the environment. As a result, certain roads can be overloaded with hazmat vehicles, which results in inequity in spatial distribution of risk.

The global route planning problem focuses on the routing decision related to multiple origin–destination pairs and multiple hazmats. This is a government problem, which is to determine the best allowable routes and lanes for certain time periods for hazmat carriers with the

TABLE 4.4 Comparison of Various Local Route and Schedule Planning Problems

Problem	En Route Decision	Real-time Update
A priori optimization	×, Routes defined before travel	×
Adaptive route selection	√, Based on historical data	×
Adaptive route selection with real-time updates	√	√

consideration of both the equity in the spatial distribution of risk and the total risk involved. The problem can be classified into

- The global route planning problem with equity
- Hazmat transportation network design

The global route planning problem with equity determines routes that minimize total risk subject to equity constraints. This problem is still a single-level optimization problem but can be extended to consider multiple objectives.

The hazmat transportation network design problem is to choose road segments that should be closed for hazmat transportation to minimize total risk, given an existing road network. This is a bilevel optimization problem. The upper level problem involves the government's decision of determining road segments to be banned for transporting hazmats for certain time periods, and the lower level problem involves carriers' problem of determining the best routes to minimize their individual transport cost.

4.7.1 Optimization Techniques for Hazmat Routing Problems

Route optimization enables decision makers to come up with better routing options with better objective values. Both single objective and multiple objective optimization approaches to the hazmat routing problem have been presented in the past. However, as discussed in Section 4.4, hazmat routing involves several stakeholders with often conflicting objectives and is inherently a multiobjective problem. Hence, the need of multiobjective modeling has been emphasized in the literature (List et al. 1991; Erkut et al. 2007).

All available models for the hazmat routing problem, whether single objective or multiobjective, can be basically classified into two categories: the shortest path models and the vehicle routing problem (VRP) models.

4.7.1.1 Shortest Path Models

A shortest path problem arises when each customer's need is satisfied by providing a single shipment. This simplifies the hazmat transport problem by eliminating the constraint of demand fulfillment. The problem in such cases reduces to finding the shortest path between a given origin and destination pair. While minimizing risk is the goal of all single objective shortest path models (Gopalan, Batta, and Karwan 1990; Kara, Erkut, and Verter 2003; Erkut and Ingolfsson 2005; Akgün et al. 2007; Carotenutoa, Giordani, and Ricciardelli 2007), risk in most studies is calculated in detailed and various aspects of risk.

For example, risk equity-based models were presented in Gopalan et al. (1990) and Carotenutoa et al. (2007). Methodologies to model the path risk components such as the link incident probability and the link consequence were analyzed in Kara et al. (2003) and Erkut and Ingolfsson (2005). The effects of weather systems were analyzed in Akgün et al. (2007).

Most multiobjective models considered both cost and risk attributes in the hazmat routing. Nonetheless, a few models basically focused on risk attributes, considering various risk factors. Both deterministic and static (Marianov and ReVelle 1998; Dell'Olmo, Gentili, and Scozzari 2005; Bonvicini and Spadoni 2008; Caramia, Giordani, and Iovanella 2010) and stochastic and dynamic (Nozick, List, and Turnquist 1997; Miller-Hooks and Mahmassani 1998; Chang, Nozick, and Turnquist 2005; Erkut and Alp 2007) multiobjective models were studied in the past.

Marianov and ReVelle (1998) linearized the bi-objective problem. They proposed assigning the weight w for the cost objective and $(1 - w)$ for the accident probability objective. The problem was then solved as a single objective problem using a well-known shortest path algorithm. They computed trade-off solutions by varying the value of w. Bonvicini and Spadoni (2008) used a similar weighted-sum-based concept to optimize the combined objective. Their model, named OPTIPATH, was basically designed for transportation risk analysis and was featured with optimization tools to optimize various strategies separately or in combination. The weighted sum approach makes the problem easier and solvable through using simple single objective optimization tools. The approach has been therefore widely applied in the hazmat transportation field.

Dell'Olmo et al. (2005) and Caramia et al. (2010) worked for obtaining a set of spatially dissimilar paths in hazmat routing so as to guarantee an equitable distribution of the risk. Both studies first used labeling algorithms for obtaining all the efficient paths between the given origin and destination nodes. The spatial dissimilarity indexes of the efficient paths in both studies were determined based on the intersection of the impact zones created around the paths. The former study used the indexes to find dissimilar paths among the efficient paths using the p-dispersion method. On the other hand, the latter study first used the k-means algorithm to categorize all efficient paths to k different classes based on the dissimilarity of their objective values (length, time, and risk) and then obtained the final dissimilar paths by choosing one path from each class while also considering for the spatial dissimilarity. In general, all stochastic and dynamic shortest path studies in hazmat routing are based on either the mean risk or stochastic dominance approaches. Moreover, the solution algorithms in many studies (Miller-Hooks and Mahmassani 1998; Chang et al. 2005) are label based.

4.7.1.2 Vehicle Routing Problem Models

A VRP is an extension to the shortest path problem in which shipments to a set of $\hat{N}$ geographically located customers are solved as a single problem. The VRP has an additional dimension of capacity, because of which it is often called the capacitated vehicle routing problem (CVRP). The problem is to determine optimal routes for k vehicles stationed at the depot to satisfy demand D_i of each customer in the set. At each customer location, there exists a constraint that the sum of demands along a route shall be less than the vehicle capacity (W_k). This constraint is known as the capacity constraint. In many real-life situations, a number of side constraints exist, which leads to variants of the VRP. The constraint of time windows at the customer locations leads to the variant called the vehicle routing and scheduling problems with time windows (VRPTW). This constraint enforces that the start of service at each customer location i should be within its prespecified time window $[b_i, e_i]$, where b_i specifies the earliest possible service time and e_i is the latest possible service time at the concerned customer location.

Compared to the shortest path problem, the VRP has been much less studied in the hazmat transportation field. Tarantilis and Kiranoudis (2001) and Pradhananga, Hanaoka, and Sattayaprasert (2011) have conducted the single objective studies toward this direction, and Zografos and Androutsopoulos (2004) and Pradhananga, Taniguchi, and Yamada (2009) have conducted multiobjective studies in this direction. Tarantilis and Kiranoudis (2001) proposed a risk-based technique to solve the CVRP in hazmat distribution. Routing decisions were made in a new space, which they called risk space. Contours of different risk levels (low, medium, and high) were determined around geographical population points. The length of each arc in the real network space was projected in the risk space. Using these projected lengths, the shortest path from each customer location to the other customer locations in the network was obtained using the well-known single objective shortest path algorithm called Dijkstra's algorithm. A single objective CVRP was then solved to reduce the total projected lengths for optimal routing. A metaheuristic algorithm called the list-based threshold accepting (LBTA) algorithm, which belongs to the class of threshold-accepting-based algorithms (Dueck and Scheuer 1990), was used to solve the problem.

Pradhananga et al. (2011) applied single objective genetic algorithm (GA) to solve a typical hazmat routing problem in Thailand. In addition to cost terms relating to the fixed cost of vehicle use, variable operation cost, and delay penalties, additional cost terms relating to traveled distance and risk exposure were also included in the objective function.

The exposures of risk were considered for both population and facilities in the area. The routing problem was solved for a single, cost-based objective. Weights could be defined for each cost term. However, the study assigned a value of 1 to all these weights without any field analysis. For path choice between customer locations, time-based shortest paths between the customers were used.

Considering hard time windows constraints, Zografos and Androutsopoulos (2004) proposed a multiobjective VRPTW model for hazmat transportation. A bi-objective model was proposed that was then transferred to a single objective model using the weighing approach. They employed an insertion-based heuristic for determining optimal routing and Dijkstra's shortest path algorithm between customer locations. The model and the heuristic algorithm were further extended in Zografos and Androutsopoulos (2008) for developing a geographic information system (GIS)-based decision support system for the integrated hazardous materials routing and emergency response decisions.

Most VRP studies in hazmat transportation were solved using the weight-based approach because this approach allows the problem to be solved easily. The Pareto-based approach is complex but it can provide the decision makers a clear view of the trade-off occurring between various objectives. Pradhananga et al. (2009) presented a Pareto-based approach to solve the VRP in the hazmat routing problem. A metaheuristic based on a multiobjective ant colony system was used to solve the problem. The algorithm returned a set of routing solutions that approximate the frontier of the Pareto optimal solutions based on total scheduled travel time and total risk of the whole transportation process.

4.8 COMBINED HAZMAT FACILITY LOCATION AND ROUTING PROBLEM

The locations of hazmat facilities determine the origin and destination of a hazmat transportation route. Hence, to reduce the total risk in a region, the hazmat facility locations must be chosen carefully. This decision can be considered with the routing decision simultaneously. The related problem is called the combined hazmat facility location and routing problem, which is to find the optimal number, capacity, and location of each hazmat facility as well as the associated optimal set of routes to be used in serving customers. This problem can be extended to capture shipping schedules, less-than-full truck loads, and other considerations in routing problems.

4.9 CONCLUDING REMARKS

This chapter discusses several dimensions of hazmat transportation, including

- The cause, the significance, and the mode of hazmat shipment
- The causes, the consequences, and the probability of the occurrence of hazmat shipment accidents
- The parties involved and their distinct objectives
- Qualitative and quantitative risk assessments
- Risk management and its four key elements: regulations, compliance, exemptions, and mitigation
- Route and schedule planning
- The relationship between the hazmat facility location decision and routing decision

Although no special examples and statistics related to Asian megacities are given, the discussion on these dimensions is believed to be still valid in the context of Asian megacities, because hazmats are required in our daily life. While this chapter serves as an introduction to students on hazmat shipment, there is another good introductory article prepared by Hancock (2001) available. If students are interested in obtaining more useful and detailed information on hazmat shipment, it is recommended that they read Erkut et al. (2007) first. Researchers who are beginners in this area will also find Erkut et al. useful as it contains a recent review and points out many future research directions. To this end, this chapter is prepared for students and researchers who are new to this area. It is therefore not surprising that most of the contents of this chapter are from other sources, including the references mentioned before.

REFERENCES

Akgün, V., Parekh, A., Batta, R., and Rump, C. M. (2007). Routing of a hazmat truck in the presence of weather systems. *Computers and Operation Research* 34 (5): 1351–1373.

Bonvicini, S., and Spadoni, G. (2008). A hazmat multi-commodity routing model satisfying risk criteria: A case study. *Journal of Loss Prevention in the Process Industries* 21 (4): 345–358.

Caramia, M., Giordani, S., and Iovanella, A. (2010). On the selection of k routes in multiobjective hazmat route planning. *IMA Journal of Management Mathematics* 21 (3): 239–251.

Carotenutoa, P., Giordani, S., and Ricciardelli, S. (2007). Finding minimum and equitable risk routes for hazmat shipments. *Computers and Operations Research* 34 (5): 1304–1327.

Chang, T. S., Nozick, L. K., and Turnquist, M. A. (2005). Multi-objective path finding in stochastic dynamic networks, with application to routing hazardous materials shipments. *Transportation Science* 39 (3): 383–399.

Dell'Olmo, P., Gentili, M., and Scozzari, A. (2005). On finding dissimilar Pareto optimal paths. *European Journal of Operation Research* 162 (1): 70–82.

Erkut, E., and Alp, O. (2007). Integrated routing and scheduling of hazmat trucks with stops en route. *Transportation Science* 41 (1): 107–122.

Erkut, E., and Ingolfsson, A. (2005). Transport risk models for hazardous materials: Revisited. *Operations Research Letters* 33 (1): 81–89.

Erkut, E., Tjandra, S., and Verter, V. (2007). Hazardous materials transportation. In *Handbooks in operations research and management science,* vol. 14, ed. C. Barnhart and G. Laporte, chap. 9, 539–621. London: Elsevier.

Federal Motor Carrier Safety Administration (2001). Comparative risks of hazardous materials and non-hazardous materials truck shipment accidents/incidents. Washington, DC.

Gopalan, R., Batta, R., and Karwan, M. H. (1990). The equity constrained shortest path problem. *Computers and Operations Research* 17 (3): 297–307.

Hancock, K. L. (2001). Dangerous goods. In *Handbook of logistics and supply chain management,* ed. A. M. Brewer et al., chap. 31, pp. 469–479. New York: Elsevier.

ICF Consulting (2000). Risk management framework for hazardous materials transportation. Submitted to Research and Special Programs Administration, US Department of Transportation.

Kara, B. Y., Erkut, E., and Verter, V. (2003). Accurate calculation of hazardous materials transport risks. *Operations Research Letters* 31 (4): 285–292.

Keller, J. J., and Associates, Inc. (2001). *Hazardous materials compliance pocketbook.* Neenah, WI: J. J. Keller and Associates, Inc.

List, G. F., Mirchandani, P. B., Turnquist, M. A., and Zografos, K. G. (1991). Modeling and analysis for hazardous materials transportation: risk analysis, routing/scheduling and facility location. *Transportation Science* 25 (2): 100–114.

Marianov, V., and ReVelle, C. (1998). Linear, non-approximated models for optimal routing in hazardous environments. *Journal of Operation Research Society* 49 (2): 157–164.

Miller-Hooks, E. D., and Mahmassani, H. S. (1998). Optimal routing of hazardous materials in stochastic, time-varying transportation networks. *Transportation Research Record* 1645: 143–151.

Nozick, L. K., List, G. F., and Turnquist, M. A. (1997). Integrated routing and scheduling in hazardous materials transportation. *Transportation Science* 31 (3): 200–215.

Pradhananga, R., Hanaoka, S., and Sattayaprasert, W. (2011). Optimization model for hazardous material transport routing in Thailand. *International Journal of Logistics Systems and Management* 9 (1): 22–42.

Pradhananga, R., Taniguchi, E., and Yamada, T. (2009). Minimizing exposure risk and travel times of hazardous material transportation in urban area. *Infrastructure Planning Review* 26 (4): 689–701.

Tarantilis, C. D., and Kiranoudis, C. T. (2001). Using the vehicle routing problem for the transportation of hazardous materials. *Operational Research—An International Journal* 1 (1): 67–78.

Transport Canada (2004). On the move—Keeping Canadians safe. Available at http://www.tc.gc.ca/Publications/TP14217e/onthemove-e.htm (accessed on August 28, 2009).

——— (2008). Transportation in Canada 2008: An overview addendum. http://www.tc.gc.ca/policy/report/aca/anre2008/pdf/addendum.pdf (accessed on August 28, 2009).

United Nations (2007). UN recommendations on the transport of dangerous goods: Model regulations. 15th revised ed. New York: United Nations.

US DOT (1998a). Hazardous materials shipments. Office of Hazardous Materials Safety, Research and Special Programs Administration, US Department of Transportation, Washington, DC.

——— (1998b). Risk based decision making in the hazardous materials safety program. Office of Hazardous Materials Safety, Research and Special Programs Administration, US Department of Transportation, Washington, DC. http://www.phmsa.dot.gov/staticfiles/PHMSA/DownloadableFiles/Files/riskprog.pdf (accessed on August 23, 2009).

——— (2004a). United States: 2002 Hazardous materials. Bureau of Transportation Statistics and US Census Bureau, Washington, DC. http://www.bts.gov/publications/commodity_flow_survey/2002/hazardous_materials/pdf/entire.pdf (accessed on August 28, 2009).

——— (2008a). Commodity summary by incidents for calendar year 2008. Printed on August 28, 2009. Pipeline and Hazardous Materials Safety Administration, US Department of Transportation, Washington, DC. http://www.phmsa.dot.gov/staticfiles/PHMSA/DownloadableFiles/Files/2008comdrank.pdf (accessed on August 28, 2009).

——— (2008b). Transportation statistics annual report. Research and Innovative Technology Administration. Bureau of Transportation Statistics, US Department of Transportation, Washington DC.

http://www.bts.gov/publications/transportation_statistics_ annual_ report/2008/pdf/entire.pdf (accessed on August 28, 2009).
——— (2009). Ten year hazardous materials incident data. US Pipeline and Hazardous Materials Safety Administration. US Department of Transportation. http://www.phmsa.dot.gov/staticfiles/PHMSA/ DownloadableFiles/Files/tenyr.pdf (accessed on August 28, 2009).
Wikipedia (2009). Pipeline and Hazardous Materials Safety Administration. http://wapedia.mobi/en/Pipeline_and_Hazardous_ Materials_Safety_Administration (accessed on August 28, 2009).
Zografos, K. G., and Androutsopoulos, K. N. (2004). A heuristic algorithm for solving hazardous materials distribution problems. *European Journal of Operational Research* 152 (2): 507–519.
———. (2008). A decision support system for integrated hazardous materials routing and emergency response decisions. *Transportation Research Part C* 16 (6): 684–703.

CHAPTER 5

Mixed Traffic in Asian Cities

Yasuhiro Shiomi

CONTENTS

5.1	Introduction	101
5.2	The On-Road Traffic Situation in Asian Countries	102
5.3	Characteristics of Mixed-Traffic Flow	105
5.4	Design of Signalized Intersections for Mixed Traffic	107
	5.4.1 Basic Concept of Traffic Capacity at Signalized Intersections	107
	5.4.2 Criticisms of PCU Value	108
	5.4.3 Capacity Analysis Considering Approach Designs	110
	5.4.3.1 Intersection Designs	110
	5.4.3.2 Estimation Method of Traffic Capacity	112
	5.4.3.3 Optimal Approach Design	115
5.5	Road Safety Issues in Mixed Traffic	117
5.6	Conclusion and Recommendation	120
References		120

5.1 INTRODUCTION

Transportation is one of the most important infrastructures that sustain social and economic activities in urban environments. Motorization, which originated in the United States in the 1920s, has widely progressed all over the world. Modern life requires motorized transportation facilities, and developed countries such as the United States and the UK have developed roadway plans to manage increasing traffic demands. Advanced traffic management systems based on intelligent transportation systems (ITS) have also been introduced to improve traffic efficiency and safety. As a result, although many problems still exist in urban transportation systems, traffic efficiency and safety have significantly improved since the dawn of motorization. Thus, it is natural that such traffic management systems have been introduced in Asian countries with less developed motorization. However, Southeast Asian motorization differs significantly from Western motorization. In most Southeast Asian countries, motorcycles play a more important role in urban traffic than four-wheel passenger vehicles. This gap between the target of the introduced traffic systems and actual traffic conditions leads to severe traffic congestion in urban environments, affecting road safety as well. Therefore, it is necessary to develop traffic control systems suitable for the local traffic situations specific to Asian countries to ensure human security and sustainable development in urban areas.

This chapter focuses on mixed-traffic flow, including motorcycles and passenger cars, that is specific to most Southeast Asian countries. Section 5.2 gives an overview of the traffic situation in Asian countries and outlines the role of motorcycles in urban transportation modes. Section 5.3 summarizes the characteristic features of mixed-traffic flow using a traffic-flow model. Next, Section 5.4 focuses on the efficiency aspect in mixed-traffic flow. Specifically, the method of evaluating traffic capacity at signalized intersections for mixed traffic is introduced. Then, the relationship between traffic capacity and intersection design is discussed. Section 5.5 outlines the tendency of traffic accidents in Vietnam on the basis of statistics and discusses effective ways to improve traffic safety. Finally, Section 5.6 provides the conclusion and recommendations for future studies.

5.2 THE ON-ROAD TRAFFIC SITUATION IN ASIAN COUNTRIES

In the 1990s, Southeast Asian countries experienced rapid economic growth, which was referred to as the "East Asia miracle" by the World Bank in 1993. In 1960, the gross domestic product (GDP) of Southeast Asian countries was equivalent to that of developing countries in other regions. Today, the economies of Southeast Asian countries continue to grow, even during the widespread global monetary crisis. With this economic growth, passenger cars and motorcycles have become more popular in Asian countries because they offer people a more convenient life. Figures 5.1 and 5.2 show the longitudinal transition in the number of passenger cars and motorcycles in use per capita in each country. The number of passenger cars per capita in use has already passed the peak in the United States, where the number of passenger cars per capita is decreasing. In the UK, France, and Japan, the number of passenger cars per capita has peaked and become stable, while it has been rapidly increasing in Asian countries.

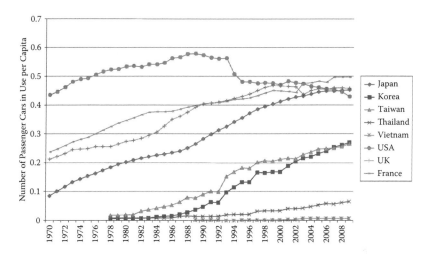

FIGURE 5.1 Longitudinal changes in passenger cars in use per capita. (World motor vehicle statistics and world statistics.)

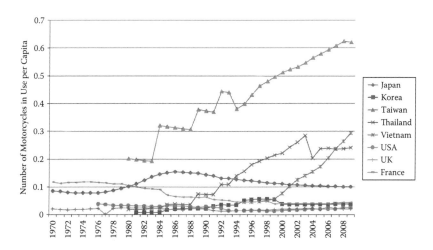

FIGURE 5.2 Longitudinal changes in motorcycles in use per capita. (World motor vehicle statistics and world statistics.)

However, the trends in the number of motorcycles in use per capita differ substantially. In most developed countries (such as the United States, the UK, France, Japan, and Korea), the number of motorcycles has gradually decreased, while in Taiwan and Thailand it continues to increase. Although the rate of increase in Taiwan has slowed recently, the numbers in Vietnam are still increasing sharply. As a consequence, the motorcycle ratio—defined as the proportion of motorcycles in use to the total number of motorized vehicles in use—is high in Vietnam, Taiwan, and Thailand, whereas it is slightly decreasing or becoming stable at lower levels in other countries, as shown in Figure 5.3. What has caused this situation?

One of the possible reasons why motorcycles have become popular in Southeast Asia is that they are less expensive than passenger cars and their maintenance costs are low as well. Figure 5.4 verifies this

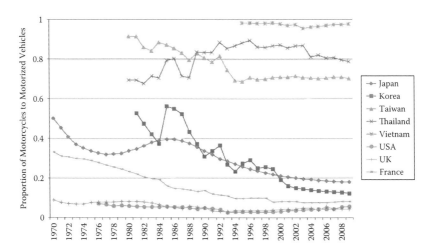

FIGURE 5.3 Proportion of motorcycles to motorized vehicles. (Source: World motor vehicle statistics and world statistics.)

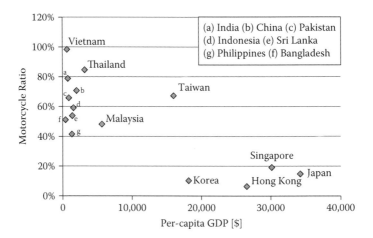

FIGURE 5.4 Relationship between the motorcycle ratio and per-capita GDP. (Source: World motor vehicle statistics and world statistics.)

conjecture by showing the relationship between the motorcycle ratio and per-capita GDP. The figure reveals that the motorcycle ratio has a strong negative relationship with GDP, although Taiwan has a significantly higher motorcycle ratio regardless of its higher per-capita GDP compared with other developing countries. Of course, there are other possible explanations:

a. Motorcycles have a higher degree of freedom compared to other transportation modes, making it easier for riders to pass through narrow alleys in urban areas.
b. Smaller, more maneuverable vehicles are less impacted by traffic congestion.
c. The hot climate in Southeast Asia is relatively more suitable to motorcycle riding than colder climates are.
d. Compact land use in urban areas eases commuting by motorcycles.

Thus, motorcycles may continue to play an important role in urban transportation in Southeast Asia until the respective economies grow sufficiently for people to afford passenger cars and public transportation systems become better organized. In areas where motorcycles are dominant, traffic flows have completely different characteristics from those where passenger cars are more dominant. To improve the present situation—that is, daily traffic congestion, noise and air pollution, and high risks of traffic accidents in urban areas—new traffic management strategies suitable for mixed-traffic situations with motorcycles and passenger cars should be developed.

5.3 CHARACTERISTICS OF MIXED-TRAFFIC FLOW

Motorcycles differ from passenger cars in terms of size, speed, and movement characteristics. Thus, motorcycle-dominated traffic and mixed-traffic flows with motorcycles and passenger cars exhibit different characteristics than passenger-car-dominated traffic flows. Here are some significant features of motorcycle behavior (Lee 2008):

a. Traveling alongside another vehicle in the same lane
b. Obliquely following another vehicle owing to its small and narrow size
c. Moving through the lateral clearances between slow moving or stationary vehicles
d. Moving to the head of a queue during a red signal
e. Swerving or weaving
f. Tailgating more closely than car drivers

Several models capturing these behavioral features have been developed from both macroscopic and microscopic perspectives. The macroscopic approach is mostly developed on the basis of the kinematic wave model, which is extended to represent the motorcycle-specific behaviors by considering areal density (Khan and Maini 1999), filtering (Powell 2000), and oblique following (Nair, Mahmassani, and Miller-Hooks 2011). Combining these extensions with a network-wide mixed-traffic simulation platform such as DynaTAIWAN (Hu et al. 2007) could help in realizing an area-wide traffic evaluation considering motorcycles, which would be useful for transportation planning, and management.

The naïve microscopic approach considers motorcycles as "just small passenger cars" moving on virtual lanes. Minh, Sano, and Matsumoto (2005a, 2005b) investigated the following and overtaking behaviors of motorcycles in mixed-traffic situations and developed motorcycle behavior models. Meng et al. (2007) developed a simulation targeting mixed-traffic flow with motorcycles and passenger cars by applying cellular automata (CA) and investigated the fundamental diagram of mixed-traffic flow as well as the relationship between the density and the number of lane changes. Matsuhashi, Hyodo, and Takahashi (2005) and Van, Schomocker, and Fujii (2009) simulated mixed-traffic flow by using VISSIM (Verkehr in Städten—Simulations). These approaches successfully represent motorcycle behavior and fundamental macroscopic features of mixed-traffic flow. However, it is still difficult to depict the behavioral features listed because they assume that motorcycles follow the traffic lanes.

Another approach to modeling microscopic motorcycle behavior involves the lane-free discipline, where it is assumed that motorcycles do not always follow lanes. Arasan and Koshy (2005) developed microscopic mixed-traffic flow simulation with the lane-free discipline. Lee (2008) developed an agent-based motorcycle behavior model considering motorcycle-specific movements that combined the longitudinal headway model, the oblique and lateral headway model, and the path choice model. Nguyen and Hanaoka (2011) developed a non-lane-based mixed-traffic flow model by applying a social force model (Helbing and Molnar 1995). The study successfully captured motorcycle behavior, but it is still necessary to model the behavior of passenger cars by considering their interaction with motorcycles to represent the entire traffic flow. Shiomi et al. (2012) developed a microscopic mixed-traffic flow model by applying a non-lane-based discrete choice approach that is often applied to depict pedestrian behaviors (Robin et al. 2009). Recently, these non-lane-based approaches have received considerable attention and successfully helped in depicting motorcycle-specific movements. Future studies are required to capture the interactions between motorcycles and passenger cars as well as represent the macroscopic characteristics of mixed-traffic flow in various conditions.

5.4 DESIGN OF SIGNALIZED INTERSECTIONS FOR MIXED TRAFFIC

This section focuses on the efficiency aspect of mixed traffic. Traffic efficiency is measured in terms of traffic capacity. In urban areas, signalized intersections can be treated as bottlenecks in terms of traffic flow. This section introduces a method to estimate the traffic capacity at signalized intersections.

5.4.1 Basic Concept of Traffic Capacity at Signalized Intersections

Traffic capacity at signalized intersections is defined as the maximum number of vehicles that can pass a given cross section during a specified effective green-light period under prevailing roadway, traffic, and control conditions. Under this concept, it is assumed that there is no influence from downstream traffic operation, such as spill-back queues into the analysis point. Traffic capacity is usually measured in passenger car unit (PCU). Basically, traffic capacity is derived by multiplying the adjustment factor parameters by the basic capacity as follows:

$$s = s_0 N \prod_i f_i, \qquad (5.1)$$

where s_0 is the basic traffic capacity, N is the number of lanes, and f_i is the adjustment factor.

In the *Highway Capacity Manual* (National Research Council 2000), lane width, heavy vehicles and grade, parking, bus blockage, area type, lane utilization, right and left turns, and pedestrians and bicycles are treated as the adjustment factors for traffic capacity at signalized intersections. The manual does not consider motorcycles because it was developed in the United States, where the motorcycle ratio is low. In contrast, in the Japanese *Traffic Engineering Handbook*, motorcycles are considered as one of the adjustment factors. Given the passenger car unit of motorcycles U_{MC} in Table 5.1, the motorcycle adjustment factor f_{MC} is calculated by Equation (5.2).

$$f_{MC} = \frac{100}{100 + \%MC \times (U_{MC} - 1)}, \qquad (5.2)$$

where %MC is the percentage of motorcycles per traffic volume.

TABLE 5.1 Passenger Car Units of Motorcycles and Bicycles

Roadway Section	PCU of a Motorcycle
Rural highway	0.75
Urban highway	0.5
Signalized intersection	0.33

FIGURE 5.5 High motorcycle ratio (afternoon-rush period in Hanoi).

However, the approach represented by Equation (5.2) indirectly assumes that no more than one motorcycle exists in one headway distance. This assumption is clearly invalid in cases where the motorcycle ratio is as high as that shown in Figure 5.5.

5.4.2 Criticisms of PCU Value

Because the PCU value is one of the most significant parameters used to determine traffic capacity, considerable attention has been paid to PCU since the early stages of mixed-traffic research. Fan (1990) investigated the passenger car equivalent (PCE) of motorcycles in Singapore and determined that the PCE values differ among countries. Rongviriyapanich and Suppattrakul (2005) showed that the PCE values varied in accordance with the motorcycle ratio. Nakatsuji et al. (2001) revealed that the impact of motorcycles on traffic flow was significantly influenced by the motorcycle's driving position in a lane and proposed a traffic-capacity estimation method at signalized intersections. Minh et al. (2005a) proposed the motorcycle equivalent unit (MUC), which indicates the number of motorcycles equivalent to other transportation modes such as a passenger car or large truck. Shiomi and Nishiuchi (2011) proposed that the saturation flow rate be evaluated using the vector of the number of motorcycles, s_{mc}, and passenger cars, s_{pc}, crossing a stop line during a certain period. In this case, when the proportion of motorcycles is the same, the distance from the origin $\sqrt{s_{mc}^2 + s_{pc}^2}$ indicates the efficiency of traffic flow. Comparisons between Taipei and Bangkok, where traffic flow is mixed but not dominated by motorcycles (see Figure 5.6), and Hanoi and Phnom Penh, where traffic flow is dominated by motorcycles (see Figure 5.7), reveal that PCU and MCU values can differ depending on the traffic situation. Even if the ratio of motorcycles is almost equal, the traffic efficiency is significantly lower in a traffic flow dominated by motorcycles (see Figure 5.8).

The recent studies mentioned before could overcome the invalid assumption in Equation (5.2). However, few studies have considered approach designs at intersections. Motorcycle behavior and traffic efficiency at signalized intersections should be strongly influenced by intersection designs. Hereafter, the method to estimate traffic capacity considering the approach designs at intersections is introduced.

FIGURE 5.6   Traffic flow in Taipei.

FIGURE 5.7 Traffic flow in Hanoi.

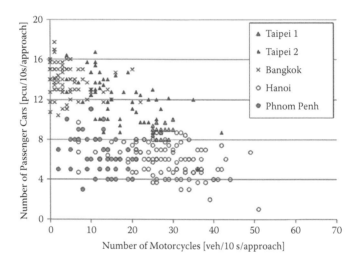

FIGURE 5.8 Scatter plot of number of motorcycles versus passenger cars in 10 s saturation flow. (Shiomi, Y., and Nishiuchi, H., 2011, *Journal of the East Asia Society for Transportation Studies* 9: 1644–1659.)

5.4.3 Capacity Analysis Considering Approach Designs

5.4.3.1 Intersection Designs

This section focuses on three types of approach designs at intersections. Figures 5.9–5.11 show the standby positions of motorcycles and passenger cars at the beginning of the green signal. Figure 5.9 shows a

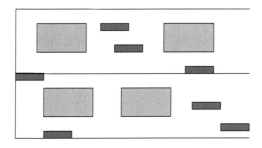

FIGURE 5.9 (See color insert.)   Do nothing.

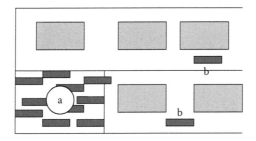

FIGURE 5.10 (See color insert.)   Waiting area.

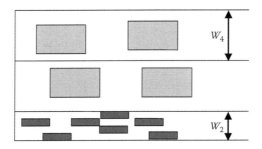

FIGURE 5.11 (See color insert.)   Motorcycle lane.

"do nothing" intersection, where all motorcycles flow with the passenger cars. Figure 5.10 shows a "waiting area" for motorcycles just behind the stop line. Motorcycles in this intersection can be divided into two types: those in the waiting area at the beginning of a green light (shown as (a) in Figure 5.10) and those flowing with the passenger cars (shown as (b) in Figure 5.10). Figure 5.11 shows an exclusive lane for motorcycle traffic. These designs are adopted in practice in Taiwan (see Figure 5.12) and some other countries.

FIGURE 5.12 (See color insert.) Motorcycle-friendly intersection design (in Tainan City, Taiwan). (a) Exclusive motorcycle lane, (b) motorcycle waiting area during red signal, (c) waiting area for two-stage left turn.)

For motorcycles that move with other traffic, their effect on the saturation flow differs in accordance with the motorcycle's position relative to passenger cars. Thus, the motorcycle's driving positions can be divided into two categories: at the center of a lane and on the side of a lane. The motorcycles belonging to these categories are called "center motorcycles" and "side motorcycles," respectively. Compared with center motorcycles, side motorcycles are assumed to have less influence on the rest of the traffic flow because they can avoid and pass passenger cars. Here it is assumed that the side motorcycles have no influence on the traffic flow, which was validated by Yoshii, Shiomi, and Kitamura (2004).

5.4.3.2 Estimation Method of Traffic Capacity

In this section, the traffic capacities at each type of intersection are investigated. For convenience, suppose that the traffic flow consists of motorcycles and passenger cars and that all of them are traveling straight.

a. Do nothing (DN): All motorcycles move with passenger cars. Under the assumption that the side motorcycles do not affect the saturation flow, the traffic capacity of the intersection Q_1 (pcu/lane/h) can be evaluated as follows:

$$Q_1 = \frac{3600 \cdot (1 - P_M)}{t_S \cdot (1 - P_M) + t_M \cdot P_M \cdot (1 - r_{side})}, \qquad (5.3)$$

where
P_M is the motorcycle ratio (the number of motorcycles divided by the number of all vehicles)
t_S is the average time headway of passenger cars in saturated flow (s)
t_M is the increased time headway by a center motorcycle (s/motorcycle)
r_{side} is the side motorcycle ratio (the number of side motorcycles divided by the number of all motorcycles)

It is assumed that t_S and t_M are constants specific to each intersection because they are strongly influenced by the lane width. Moreover, it is assumed that r_{side} is a constant unique to each intersection, although it is affected not only by the lane width but also by the green ratio (green length divided by cycle length) and motorcycle ratio.

b. Waiting area (WA): Motorcycles in this intersection are divided into two types, **a** and **b**, as shown in Figure 5.10 The former impedes the passenger car flow because it obstructs passenger cars waiting behind in the waiting area. As a result, the motorcycles cause start-up loss time to increase. Considering this influence, the traffic capacity of the intersection, which consists of n lanes with one of them having a waiting area for motorcycles, Q_2 (pcu/lane/h) can be evaluated as follows:

$$Q_2 = \frac{3600 \cdot (1 - P_M)}{t_S \cdot (1 - P_M) + t_M \cdot P_{MF}} \cdot \left(1 - \frac{\alpha \cdot N_{MW} + \beta}{n \cdot C \cdot g}\right), \quad (5.4)$$

$$P_{MF} = (1 - r_{side}) \cdot \{1 - k \cdot (1 - g)\} \cdot P_M, \quad (5.5)$$

$$N_{MW} = k \cdot (1 - g) \cdot \frac{P_M}{1 - P_M} \cdot \frac{C \cdot g}{3600} \cdot Q_2 \quad (5.6)$$

where
l_M is the start-up loss time of the passenger car in the lane that has the waiting area—namely, the time interval from the beginning of the green light to the time when the first passenger car passes across the stop line (s)
C is the cycle length (s)
G is the split of the concerned approach
N_{MW} is the number of motorcycles starting from the waiting area per each cycle
α, β, κ are parameters

It is assumed that t_S, t_M, and r_{side} are constants specific to each intersection. Actually, all the motorcycles cannot reach the waiting area because they are blocked by queuing passenger cars. Therefore, parameter k is introduced, which is the proportion of motorcycles in the waiting area to the ones arriving at the intersection during the red signal.

c. Motorcycle lane (ML): The widths of the motorcycle lane (W_2) and a passenger car lane (W_4) should be determined on the basis of the motorcycle ratio. If the motorcycle ratio increases, it is necessary to assign a wider lane to motorcycles. As a result, the lane width for passenger cars should be smaller, and the saturation flow rates of each lane are required to decrease. If the share of the motorcycle lane is not set appropriately, the motorcycle or passenger car lanes are unable to maintain their saturation flow. This situation is considered suboptimal because it prevents maximum traffic flow at the intersection. Therefore, it is most efficient to determine the appropriate lane width for which queues of both motorcycles and passenger cars, which are formed during the red signal, disappear at the same time. Namely, when the proportion of the maximum motorcycle's flow rate to that of passenger cars is equal to the proportion of the demand of motorcycles to that of passenger cars, the most efficient design can be realized. Accordingly, the necessary condition for the efficient design is expressed as

$$\frac{S_M}{W_2} : \frac{t_S}{n} = \frac{1}{P_M} : \frac{1}{1-P_M}, \qquad (5.7)$$

$$W_2 + n \cdot W_4 = W, \qquad (5.8)$$

where
W is the total road width (m)
W_2 is the motorcycle lane width (m)
W_4 is the passenger car lane width (m)
n is the number of passenger car lanes
s_M is the time headway of a motorcycle in saturated flow on a 1-meter width lane (s/motorcycle · m)

The time headway of passenger cars t_S is determined by the lane width:

$$t_S = f(W_4) = \frac{t'_S}{0.24 \cdot W_4 + 0.22}, \qquad (5.9)$$

where t_S' is the basic time headway of a passenger car in saturated flow (s).

Thus, the traffic capacity of the intersection Q_3 (pcu/lane/h) is estimated in relation to the motorcycle ratio as follows:

$$Q_3 = \frac{3600}{t_S}. \tag{5.10}$$

5.4.3.3 Optimal Approach Design

This section reviews the traffic capacity of each approach. The assumed approach, which has a width of 10.5 m, is considered and the relationship between the motorcycle ratio and traffic capacity of the approach under each design is discussed. Both DN and WA use three lanes that are 3.5 m wide. The ML approach uses three lanes for passenger cars and one motorcycle lane with a minimum width requirement of 1.5 m. Let 2.0 s be the base time headway of a passenger car in saturated flow. On the basis of this and the results of the survey (Yoshii et al. 2004), each parameter value that is necessary to estimate the traffic capacity of each intersection is fixed (see Tables 5.2–5.4).

The relationship between the motorcycle ratio and traffic capacity is shown in Figure 5.13. We obtained the parameter values for DN and WA from the observation survey under a low motorcycle ratio, but the values may change significantly when the motorcycle ratio increases considerably. In this analysis, it is assumed that these parameter values become invalid when the motorcycle ratio is higher than 0.5. Thus, the traffic capacity lines for DN and WA are illustrated by dotted lines in the figure. Moreover, because the traffic capacity is affected by the split in the case of WA, the traffic capacities during splits of 0.4 and 0.6 are described in the figure.

TABLE 5.2 Parameter Set for DN

t_S [S]	t_M [S]	r_{side}
2.00	0.68	0.45

TABLE 5.3 Parameter Set for WA

t_S [S]	t_M [S]	r_{side}	C [S]	k	α	β
2.00	0.56	0.53	200.00	0.60	0.27	3.50

TABLE 5.4 Parameter Set for ML

t'_S[S]	S_M [S]
2.00	0.81

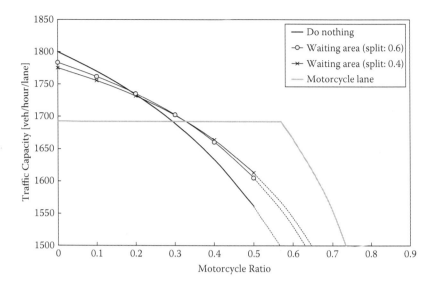

FIGURE 5.13 Traffic capacity versus motorcycle ratio.

Compared to the DN, WA has two effects on traffic capacity: One is an increasing effect because many motorcycles move in a bunch, while the other is a decreasing effect because start-up loss time increases. When the motorcycle ratio is low, the latter effect is stronger than the former, and therefore the capacity under DN is higher than that under WA. On the other hand, when the motorcycle ratio is high, the former effect becomes stronger, and the capacity under WA is higher than that under DN. Concerning the two lines representing the capacity under WA, because the share of motorcycles in the waiting area increases when the split is smaller, the capacity for a smaller split has a stronger effect than that for a larger split. Therefore, the capacity for a smaller split is greater than that for a larger split when the motorcycle ratio is low, and the capacity is less than that for a larger split when the motorcycle ratio is high. As for ML, because the minimum lane width is required for motorcycles, when the motorcycle ratio is low, the motorcycle lane does not work efficiently and the capacity is less than that of other approaches. However, when the motorcycle ratio is sufficiently high, ML works efficiently and the highest capacity can be achieved.

On the basis of these results, when the motorcycle ratio is low (about 0.25 in this study), DN is most efficient. With an increase in the ratio, WA becomes the most efficient. Finally, when the motorcycle ratio is higher than 0.35 in this study, ML becomes the most efficient. These results confirm that the most efficient intersection design varies depending on the motorcycle ratio.

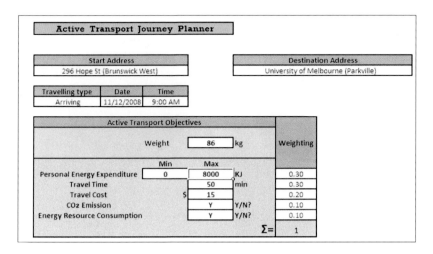

FIGURE 3.2 Active transport journey planner (input).

Recommendation Route: 5

Route Ranking	Route	Trip Description	Parking Cost ($)	Energy Expenditure (KJ)	Travel Time (min)	Travel Cost ($)	CO_2 Emission (g)	Energy Resource Consumption (MJ)	Feasibility
5	1	Home Walking Tram Walking Bus Walking University	0.00	602.40	42.82	1.53	277.10	5.59	Y
4	2	Home Walking Bus Walking Tram Walking University	0.00	502.00	45.18	1.53	308.60	10.94	Y
6	3	Home Car Walking University	7.50	25.10	13.31	11.65	1066.40	29.14	Y
1	4	Home Car Bicycle University	0.00	4062.96	18.56	1.91	481.60	13.42	Y
2	5	Home Bicycle University	0.00	7724.64	24.64	0.06	0.00	0.49	Y
3	6	Home Motorcycle University	0.00	0.00	12.40	0.74	768.80	17.36	Y

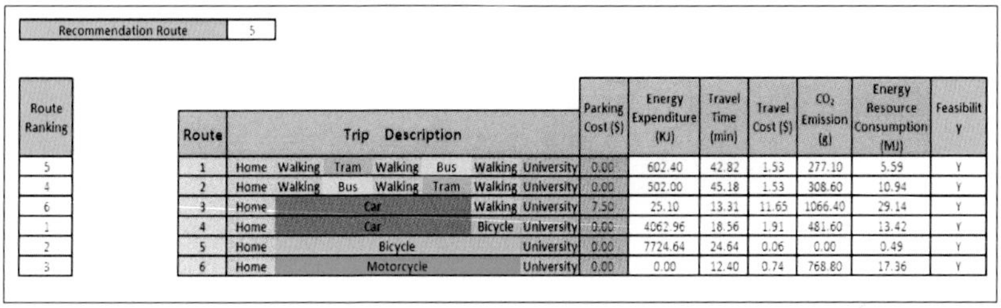

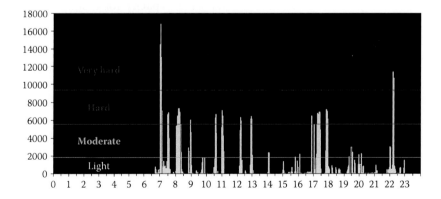

FIGURE 3.3 Active transport journey planner (output).

FIGURE 3.4 Activity count profile.

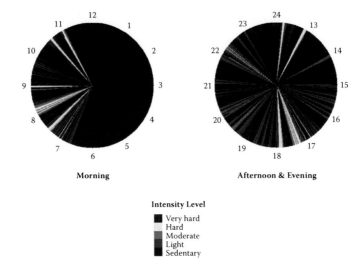

FIGURE 3.5 Intensity levels by time of day.

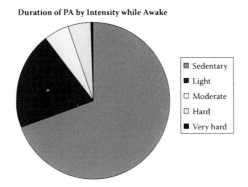

FIGURE 3.6 Duration of physical activity by intensity levels.

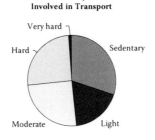

FIGURE 3.8 Duration of intensity levels for transport.

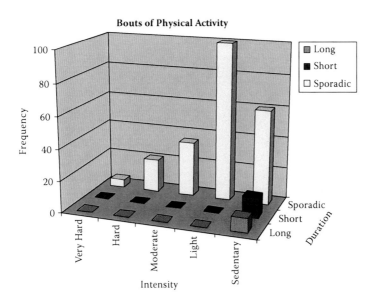

FIGURE 3.9 Bouts of physical activity by frequency, duration, and intensity.

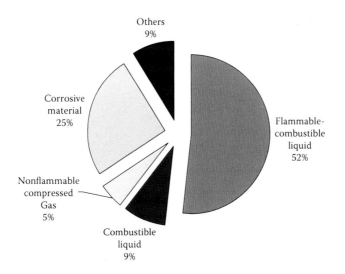

FIGURE 4.1 Accident/incident by HAZMAT class in 2008. (US DOT, 2008a, Pipeline and Hazardous Materials Safety Administration, U.S. Department of Transportation, Washington, DC. http://www.phmsa.dot.gov/staticfiles/PHMSA/DownloadableFiles/Files/2008comdrank.pdf.)

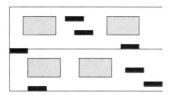

FIGURE 5.9 Do nothing.

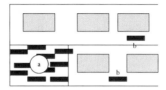

FIGURE 5.10 Waiting area.

FIGURE 5.11 Motorcycle lane.

FIGURE 5.12 Motorcycle-friendly intersection design (in Tainan City, Taiwan). (a) Exclusive motorcycle lane, (b) motorcycle waiting area during red signal, (c) waiting area for two-stage left turn.)

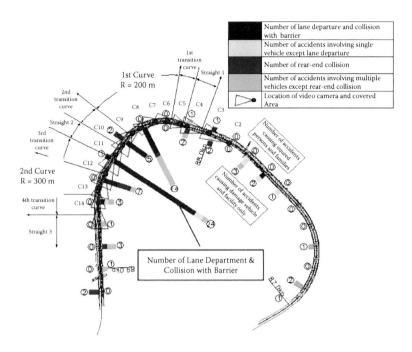

FIGURE 6.6 Location and type of accidents recorded at Nakahata Curve.

FIGURE 6.7 Example of images obtained from the video cameras.

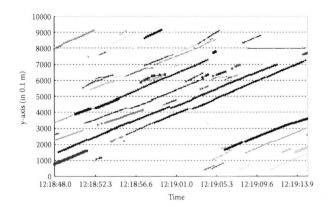

FIGURE 6.9 Examples of raw vehicle trajectory data.

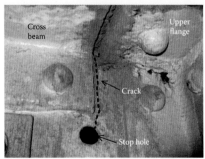

FIGURE 6.12 Fatigue cracks of a concrete slab (left) and cross beam (right) in Japan.

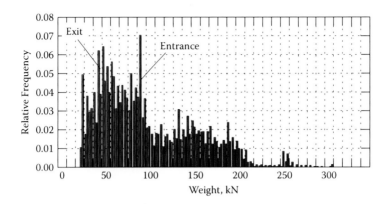

FIGURE 6.21 Relative frequency of gross weight in exit and entrance lanes.

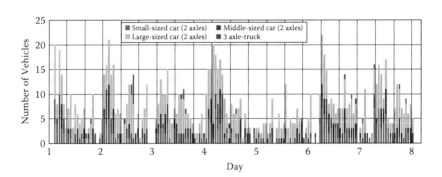

FIGURE 6.22 Time series of passing vehicles according to the vehicle type (exit lane).

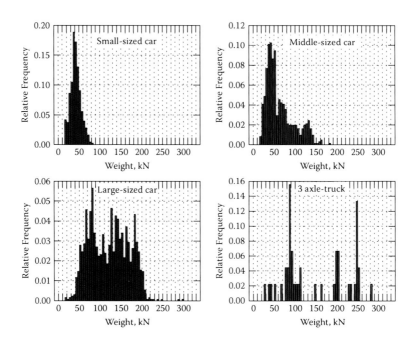

FIGURE 6.23 Relative frequency of gross weight (according to the types of vehicles).

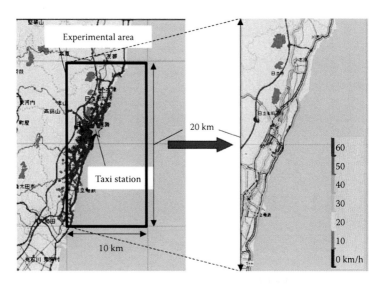

FIGURE 10.4 Probe car system experiment example in Hitachi, Japan.

5.5 ROAD SAFETY ISSUES IN MIXED TRAFFIC

Road safety is one of the most significant concerns in cities with mixed traffic. As economies grow, motorized vehicles are more commonly used in urban areas and mixed-traffic situations become more chaotic, causing more traffic accidents. As shown in Figure 5.14, in the case of Vietnam, the number of traffic accidents has increased more than three times over two decades with rapid growth in motorization. It is interesting to note that the number of accidents is almost equal to the number of injuries and fatalities, which implies that these statistics indicate only traffic accidents with fatalities and/or injuries. The actual number of traffic accidents, including those without injuries, may be significantly higher than these statistics show.

The behaviors specific to motorcycles also may cause traffic accidents. Figure 5.15 summarizes the proportion of each accident type. The figure shows that more than half of all injuries and fatalities are caused by accidents between motorcycles or between motorcycles and cars and that more than 70% of the injuries, fatalities, and accidents involved motorcycles. Because motorcycle riders are exposed and are vulnerable to traffic accidents, even minor collisions can cause motorcycle rider fatalities. Figure 5.16 illustrates this by comparing the number of fatalities per 10,000 population in Vietnam with that in Japan. Although the number of accidents per capita is higher in Japan than that in Vietnam, the fatality rate in Vietnam has become much higher than that in Japan. On December 15, 2007, it became mandatory for all riders of motorized

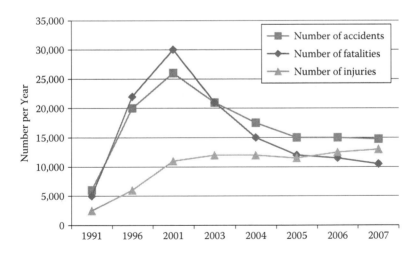

FIGURE 5.14 Number of accidents, fatalities, and injuries in Vietnam. (Source: JICA-NTSC traffic safety master plan, 2008.)

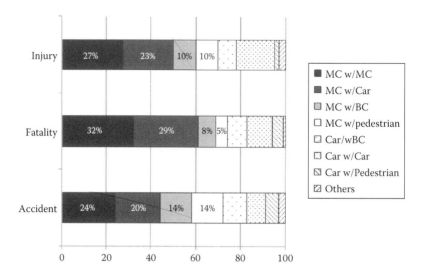

FIGURE 5.15 Breakdown of traffic accident types. (Source: JICA-NTSC traffic safety master plan, 2008.)

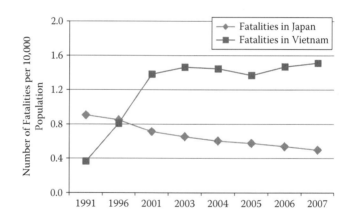

FIGURE 5.16 Number of fatalities per 10,000 population. (Source: JICA-NTSC traffic safety master plan, 2008.)

two-wheelers to wear helmets in Vietnam; this is expected to reduce the number of fatalities and severe injuries.

To mitigate traffic accidents in mixed-traffic situations, the segregation concept was proposed by Hsu (1997). In this concept, motorcycle and passenger-car traffic is segregated in terms of space and time (as shown in Figure 5.17) to prevent side-by-side collisions, which are the most

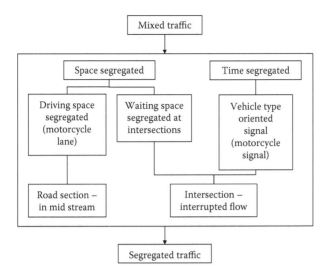

FIGURE 5.17 Segregated traffic concept. (Hsu, T. P. et al., 2003, *Journal of the Eastern Asia Society for Transportation Studies* 5: 179–193.)

TABLE 5.5 Merits and Demerits of Motorcycle Segregation

	Merits	Demerits
Two-stage left turn	Decreases traffic conflicts and traffic accidents (Hsu 2004)	Increases travel time of motorcycles
Motorcycle waiting area	Decreases traffic conflicts between motorcycles and passenger cars	Increases the start-up loss time of passenger cars
	Increases saturation flow rate of both motorcycles and passenger cars	Increases the conflicts between motorcycles
Motorcycle segregation lane	Decreases traffic conflicts between motorcycles and passenger cars (Hsu 2004)	Increases the conflicts between motorcycles
	Increases saturation flow rate and traffic speed (Hsu 2004)	Owing to speed-up, the severity of traffic accidents increases
		Decreases traffic capacity for passenger cars

common type of traffic accidents involving motorcycles (Hsu 2004). In practice, the number of traffic accidents decreased in sites where motorcycle-exclusive lanes were installed (Hsu 2004). However, these segregation schemes may have both advantages and disadvantages in terms of safety and efficiency (Table 5.5). It is important to develop an assessment method to determine the safety and efficiency of motorcycle segregation.

5.6 CONCLUSION AND RECOMMENDATION

This chapter described the mixed-traffic situation in Asian countries, covering vehicle ownership and economic conditions, mixed-traffic flow modeling, saturation-flow analysis, and approach designs at signalized intersections and traffic safety issues. Since the 2000s, because of their rapid economic growth, Asian countries have received considerable attention, and the number of research papers about motorcycles and mixed-traffic issues has increased. However, the established methodologies and information associated with motorcycle traffic are still limited compared with those associated with passenger vehicles. One of the difficulties in the motorcycle research is in its observation. In the case of passenger vehicles (i.e., lane-based traffic flow), cross-sectional traffic observation is useful (e.g., loop detectors, ultrasonic detectors, and AVI [automated vehicle identification]).

However, motorcycle traffic flow is not lane based, so it is difficult to identify each motorcycle using fixed-point detectors. Nowadays, image-processing techniques and mobile phones with GPS (global positioning system) sensors and acceleration meters are commonly used in this area. These "new" observation tools and data could be a breakthrough in mixed-traffic and motorcycle research. It is strongly recommended that future studies in this regard establish a highway capacity manual for Asian countries by analyzing the socioeconomics of Asian transportation, traffic engineering for efficiency and safety, and the psychology of drivers and riders in mixed-traffic situations.

REFERENCES

Arasan, V., and Koshy, R. (2005). Methodology for modeling highly heterogeneous traffic flow. *ASCE Journal of Transportation Engineering* 131 (7): 544–551.

Fan, H. S. L. (1990). Passenger car equivalents for vehicles on Singapore expressways. *Transportation Research* 24 (5): 391–396.

Helbing, D., and Molnar, P. (1995). Social force model for pedestrian dynamics. *Physical Review E* 51 (5): 4282–4286.

Hsu, T. P. (1997). Development of motorcycle traffic engineering and separated traffic flow concept. *Urban Transportation* 91: 41–49.

——— (2004). Segregated flow concept for motorcycle traffic. *IATSS Review* 29 (3): 32–42.

Hsu, T. P., Sadullah, E. A. F. M., and Dao, I. N. X. (2003). A comparison study on motorcycle traffic flow development of Taiwan, Malaysia and Vietnam. *Journal of the Eastern Asia Society for Transportation Studies* 5: 179–193.

Hu, T. Y., Liao, T. Y., Chen, L. W., Huang, Y. K., and Chiang, M. L. (2007). Dynamic simulation-assignment model (DynaTAIWAN) under mixed traffic flows for ITS Applications. *TRB 86th Annual Meeting Compendium of Papers* CD-ROM.

Japan Automobile Manufactures Association, Inc. (1973–2012). World motor vehicle statistics Tokyo, in Japanese.

Japan Society of Traffic Engineers (2001). *Traffic engineering handbook.* CD-ROM, in Japanese.

Japan Statistical Association (1994–2012). World's statistics, in Japanese.

JICA-NTSC (2009). Final report of the survey for planning the master plan of road traffic safety in Vietnam.

Khan, S., and Maini, P. (1999). Modeling heterogeneous traffic flow. *Transportation Research Record* 1678: 234–241.

Lee, T. C. (2008). An agent-based model to simulate motorcycle behavior in mixed traffic flow. PhD dissertation, Imperial College, London.

Matsuhashi, N., Hyodo, T., and Takahashi, Y. (2005). Image processing analysis of motorcycle oriented mixed traffic flow in Vietnam. *Proceedings of the Eastern Asia Society for Transportation Studies* 5: 929–944.

Meng, J. P., Dai, S., Dong, L., and Zhang, J. (2007). Cellular automaton model for mixed traffic flow with motorcycles. *Physica A* 380: 470–480.

Minh, C. C., Sano, K., and Matsumoto, S. (2005a). The speed, flow and headway analyses of motorcycle traffic. *Journal of the Eastern Asia Society for Transportation Studies* 6:1496–1508.

——— (2005b). Characteristics of passing and paired riding maneuvers of motorcycle. *Journal of the Eastern Asia Society for Transportation Studies* 6: 186–197.

Nair, R., Mahmassani, H. S., and Miller-Hooks, E. (2011). A porous flow approach to modeling heterogeneous traffic in disordered systems. *Transportation Research Part B* 45 (9): 1331–1345.

Nakatsuji, T., Hai, N. G., Taweesilp, S., and Tanaboriboon, Y. (2001). Effects of motorcycle on capacity of signalized intersections. *Proceedings of Infrastructure Planning* 18 (5): 935–942.

National Research Council (2000). *Highway capacity manual 2000.* Transportation Research Board special report 209. Washington, DC: Transportation Research Board.

Nguyen, L. X., and Hanaoka, S. (2011). An application of social force approach for motorcycle dynamics. *Proceedings of the Eastern Asia Society for Transportation Studies* 8: 319–326.

Powell, M. (2000). A model to represent motorcycle behavior at signalized intersections incorporating an amended first order macroscopic approach. *Transportation Research Part A* 34: 497–514.

Robin, T., Antonini, G., Bierlaire, M., and Cruz, J. (2009). Specification, estimation and validation of a pedestrian walking behavior model. *Transportation Research Part B* 43 (1): 36–56.

Rongviriyapanich, T., and Suppattrakul, C. (2005). Effects of motorcycle on traffic operations on arterial streets. *Journal of the East Asia Society for Transportation Studies* 6: 137–146.

Shiomi, Y., Hanamori, T., Uno, N., and Shimamoto, H. (2012). Modeling traffic flow dominated by motorcycles based on discrete choice approach. *Proceedings of 1st LATSIS Conference.*

Shiomi, Y., and Nishiuchi, H. (2011). Evaluation of spatial motorcycle segregation at isolated signalized intersections considering traffic flow conditions. *Journal of the East Asia Society for Transportation Studies* 9: 1644–1659.

Van, H. T., Schomocker, J. D., and Fujii, S. (2009). Upgrading from motorbikes to cars: Simulation of current and future traffic conditions in Ho Chi Minh City. *Proceedings of the Eastern Asia Society for Transportation Studies* 8, pp. 335–349.

Yoshii, T., Shiomi, Y., and Kitamura, R. (2004). Capacity analysis of mixed traffic flow with motorcycles. *IATSS Review* 29 (3): 178–187, in Japanese.

CHAPTER 6

Road Safety

*Nobuhiro Uno, Yasunobu Oshima,
and Russell G. Thompson*

CONTENTS

6.1 Introduction to Road Safety 124
    6.1.1 The Traffic System 124
        6.1.1.1 Human 125
        6.1.1.2 Vehicle 125
        6.1.1.3 Road 125
    6.1.2 Crash Statistics 126
    6.1.3 Data Sources 127
        6.1.3.1 Exposure 127
6.2 Hazardous Road Locations (HRLs) 128
    6.2.1 Identification of Hazardous Road Locations 128
    6.2.2 HRL Programs 129
    6.2.3 Diagnosis of Road Crash Problems 129
        6.2.3.1 In-Office Analysis 129
    6.2.4 Prevention of Road Crashes 130
        6.2.4.1 Implementation 130
        6.2.4.2 Road Crash Costs 130
    6.2.5 Evaluation of Effectiveness 130
    6.2.6 Design of Evaluation Studies 131
6.3 Road Safety Audits 132
    6.3.1 What Are Road Safety Audits? 132
    6.3.2 Steps in Conducting a Road Safety Audit 133
    6.3.3 Safer Road Design 134
        6.3.3.1 Safe Road Environments 134
        6.3.3.2 Safe Road Design Principles 134
6.4 Outlines of Traffic Safety Issues 135
    6.4.1 Trends and Causes of Accidents 135
    6.4.2 Difficulty of Traffic Safety Studies 136
6.5 ITS for Improvement in Traffic Safety 137

6.5.1 Important Viewpoint for Countermeasures for
  Improving Traffic Safety   137
 6.5.2 Possible Contribution of ITS to Improving Traffic Safety   138
6.6 A Review of Traffic Conflict Analyses   139
6.7 Utilization of Video Image Data for Conflict Analysis   141
 6.7.1 Outlines   141
 6.7.2 An Example of Conflict Analysis Using Video Image
  Data   142
   6.7.2.1 Outline of Studied Section   142
   6.7.2.2 Traffic Accidents Occurred at Nakahata Curve   143
   6.7.2.3 Data Used for Analysis   144
   6.7.2.4 An Example of Evaluating the Potential
    Danger of Lane Departure   145
   6.7.2.5 Analysis Based on Estimated Centrifugal
    Acceleration   146
6.8 Structural Safety   147
 6.8.1 Overloaded Vehicles in Asian Countries   148
 6.8.2 Impacts of Heavy Vehicles on Structures   148
 6.8.3 Bridge Weigh-in-Motion   150
   6.8.3.1 Estimation by the Strain Responses of Main
    Girders   151
   6.8.3.2 Estimation by Responses of the Second
    Members   155
   6.8.3.3 Estimation by Crack Response of Reinforced
    Concrete Slab   157
   6.8.3.4 Error Factors of BWIM   157
 6.8.4 Case Study in Bangkok, Thailand   158
References   163

6.1 INTRODUCTION TO ROAD SAFETY

Although the road system offers a high degree of mobility in Asian cities, a large amount of personal trauma is experienced through crashes. The term "crash" is preferred to "accident" since accident implies accidental and these events are unavoidable and nonpreventable. Crashes are considered a failure of the system that can be eliminated through good engineering.

6.1.1 The Traffic System

The traffic system comprises three major components: humans, vehicles, and roads (Figure 6.1). Both human (drivers or controllers) and vehicle

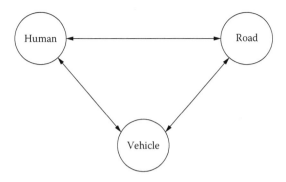

FIGURE 6.1 The traffic system.

(moving parts) characteristics are essential elements for road design and management. The properties of both elements should heavily influence the design and safety of roads.

The road element is the primary concern of the traffic engineer, but familiarity with the other components is essential. All three elements must be fully compatible for the system to operate without failure (e.g., crashes and congestion).

6.1.1.1 Human

Humans control the vehicle, taken as given (static), but there must be detailed knowledge of the characteristics (e.g., vision, memory, reaction time, judgment, expectancy, information processing, and sensing), which may differ by age, gender, and experience. These lead to rules (e.g., drivers should be confronted only with one simple decision at one time). Human factors should influence design geometry, alignment, signs, signals, etc. Personal characteristics like eye height, vision range, etc., need to be identified. Generally, we design for the 85th percentile, to ensure both safety and economic performance.

6.1.1.2 Vehicle

Vehicles operate in road environments and their characteristics need to considered to ensure that they can safely operate. Vehicle dimensions such as length, width, and weight need to be identified and considered for safe road design and management. Performance of vehicles, such as acceleration, cornering, and braking, also needs to be considered.

6.1.1.3 Road

The road environment is the main concern of transport engineers. Road design and operational management are important areas that have a large influence on road safety (Underwood 2006; Ruediger and Mailaender 1999; Ogden 1996). Although driver and vehicle factors are

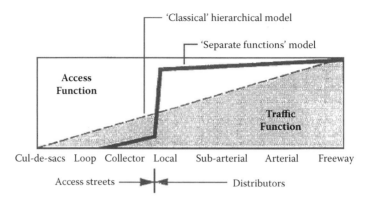

FIGURE 6.2 Functional mix of roads and streets (Source: Ogden, K. W., and Bennett, D. W., 1989, *Traffic engineering practice*, 4th ed. Clayton, Australia: Department of Civil Engineering, Monash University, Chapter 6.7.)

often identified as the cause of crashes, a significant proportion of road crashes is attributable to the road environment.

Road crashes occur when a number of adverse factors *combine* to cause a failure of the traffic system. Although the dominant factor in road crash causation is often the driver, the road environment often significantly contributes to likelihood and severity of road crashes (Lay 1985).

Although there have recently been a number initiatives to improve the performance of vehicles as well as increase the awareness of road safety and responsibility of drivers, techniques aimed at improving road conditions will be emphasized here.

It is important to determine an appropriate functional classification for roads in a network. Roads have two primary road functions: through movement and property access. However, these are conflicting! Safety problems typically occur when roads attempt to serve both functions. Contemporary road network planning emphasizes the separation of functions (Figure 6.2).

6.1.2 Crash Statistics

Road crashes are generally classified in a number ways:

- Severity level (e.g., fatality and personal injury)
- Road user groups
    - Vehicle type (e.g., passenger vehicles, trucks, etc.)

- Region (e.g., country, city, state, and local government area [LGA])
- Driver characteristics (e.g., age and experience)
- Exposure measures/rates (e.g., per population, registered vehicles, and vehicle kilometers traveled [VKT])
- Road user movement (RUM)
- Time

6.1.3 Data Sources

There are a usually a number of sources of crash statistics:

- Government departments (e.g., state road authorities and VicRoads [Roads Corporation of Victoria])
- Police departments
- Research organizations (e.g., ARRB Transport Group and Monash University Accident Research Center [MUARC])

Care must be used when comparing accident data due to different recording and reporting criteria.

There is generally a range of inputs into programs to reduce crashes, such as

- Political
- Budget constraints
- Relative effectiveness of each available countermeasure in terms of its potential for reducing the frequency or severity of crashes
- Maintaining the same level of safety while increasing the flow of traffic

Statistical analysis based on scientific principles as well as engineering judgment provides a sound basis for developing crash reduction programs.

Statistical analysis is an integral part of accident research. The rationale for safety funding is the reduction in crash frequency or severity. Countermeasure programs need to be *evaluated* rigorously to be absolutely sure that society's resources are being expended on programs that really work.

6.1.3.1 Exposure

Crash data must be compared with the experience of the nonaccident population. Exposure metrics are commonly used where the numerator is the number of accidents and the denominator is the measure of the population exposure—for example, volume or length of road segment.

There is usually a wide variety of data sources relating to crashes, including:

- Police department accident reports
- Coroner's inquest reports
- Road authority reports
- Insurance company records
- Tow-truck operators records
- Research studies

In many countries, police departments are responsible for investigating crashes. Although reporting criteria and recording procedures differ between jurisdictions, details of the location, persons involved, time, RUM, and severity are usually recorded.

Data gained from the accident report form often lack the details of the road environment necessary for establishing causal relationships between road crashes and road design. It is therefore often necessary to combine road usage and road inventory data using analysis tools such as a geographic information system (GIS).

6.2 HAZARDOUS ROAD LOCATIONS (HRLS)

6.2.1 Identification of Hazardous Road Locations

Several statistics are generally used to identify elements of the road network that have a high crash frequency or severity levels (NAASRA 1988).

Black spots are clusters of road crashes occurring at precise locations identified by features of road geometry such as junctions and bends. Black sites (specific lengths of road) and black areas are other locations of high risk, which may be identified individually from accident history in terms of clusters of accidents.

The concept of exposure is important when comparing crash sites. Exposure is a measure of the opportunity for a crash to occur (e.g., traffic volume or VKT). Network elements are often analyzed by

- Crash type—classification of vehicle movement or road user involved
- Crash rate—number of crashes per unit of exposure

- Crash severity—a measure of the consequence of the crash in terms of the casualty sustained
- Casualty class—a measure of the consequence of a crash in terms of the number of injuries

6.2.2 HRL Programs

Programs aimed at reducing the crashes at specific locations generally have a number of phases, including:

a. Identification of HRLs
b. Investigation
   - Diagnosis of crash problems
   - Selection of countermeasures
c. Implementation
   - Ranking of priority for treatment
   - Preparation of design plans
   - Implementation of countermeasures
d. Evaluation of Countermeasures

Common HRL identification criteria at intersections are the number of crashes, crash rate, and severity rate. For road sections or midblocks, the number of crashes/kilometer and severity index are generally used.

6.2.3 Diagnosis of Road Crash Problems

6.2.3.1 In-Office Analysis

Collision diagrams are generally used to investigate the cause of crashes as part of the in-office analysis. This summarizes all vehicle movements involved in collisions and highlights predominant crash types and vehicle maneuvers. Other details from the crash report, such as the date and time of the crash, condition of the road, light conditions at time of the crash, geometry of the site, and type of vehicle involved, are usually analyzed.

An on-site inspection is necessary to assess road conditions and other site factors accurately. This should identify adverse features of road design and adverse environmental features.

Other factors influencing behavior at the site, such as the physical and mental condition, experience, and age of drivers should also be investigated. When selecting countermeasures, a large degree of judgment and experience is necessary.

6.2.4 Prevention of Road Crashes

Road design and traffic management techniques can reduce the potential for and the occurrence of road crashes (Underwood 2006). Geometric design features associated with urban networks such as intersection design can cause crashes. Traffic control devices are often used to improve road safety. Signs (regulatory, warning, and guide), signals, and pavement markings are often used. A number of treatments can increase protection for pedestrians, such as crossings, medians, refuge islands, overpasses, and underpasses. Street lighting also provides safety benefits for pedestrians.

There are specialized planning and design techniques that are used to design safe street systems in residential areas (Ogden 1996).

6.2.4.1 Implementation

Governments need to develop a works program to allocate funds to improve road safety. This requires individual projects to be ranked due to budgetary constraints. The aim of the road safety program is to maximize economic benefits. It is common to use standard economic evaluation criteria such as the benefit-cost ratio (BCR) or net present value (NPV) to rank projects.

6.2.4.2 Road Crash Costs

Implementation cost estimates are based on typical treatments and experience, including maintenance costs. Estimates need to be made for each countermeasure and site. Here, a crash reduction factor (the expected reduction by countermeasure by crash type) needs to be estimated and is generally based on previous performance. Road crash costs are generally tabulated as per person by severity or per vehicle. These need to be converted to per-crash cost.

6.2.5 Evaluation of Effectiveness

A process for determining whether or how effectively a road safety activity has bought about the desired result is necessary for ensuring financial accountability as well as establishing causal relationships. This process relies heavily on crash data and statistical testing.

  a. Evaluation procedures: An evaluation program should consider details of
     - The unsatisfactory situation
     - The countermeasure introduced
     - Significant changes that occurred or did not occur

b. Selection of variable to be measured: Determining the variable to be measured is important and involves consideration of what crash frequency or rate (maybe subcategory [type of crash]) should be considered as well what type of crash the particular countermeasure is expected to affect. It is best to compare statistics based on exposure and to compare a severity criterion.

There are a number of issues that can threaten the validity of evaluations, including:

- Regression to the mean effect
- Other causes operating at the same site
- Trends over time
- Data considerations
  - Crash reporting criteria
  - Coding effects
  - Biased data
  - Instability
- Crash migration effects

6.2.6 Design of Evaluation Studies

It is necessary to design a procedure for evaluating the effect of countermeasure treatments:

a. Before/after study design with randomized control group: A group of locations which are candidates for a given treatment are first randomly assigned to either a treatment group or a control group. The "predicted" level of the after measure is based on experience of the control group. Statistical techniques are used to test the significance of differences.
b. Before/after study design with comparison group: This is similar to the previous method but the groups are not assigned on a random basis. This can build a comparison group after the fact.
c. Before/after study design correcting for regression to the mean effect: This needs to estimate the regression to the mean effect to predict after conditions if no treatment was implemented.
d. Simple before/after study design: This involves comparing the performance of locations before and after treatment. However, this has several weaknesses, such as

- Regression to the mean effect not included
- Other causes at the same time (e.g., weather)
- Trends over time

Numerous statistical tests are recommended for assessing the significance of effects of treatments based on the evaluation design and the type of criterion (NAASRA 1988).

6.3 ROAD SAFETY AUDITS

6.3.1 What Are Road Safety Audits?

Road safety audits (RSAs) are a formal examination of an existing or future road or traffic project—or any project that interacts with road users—in which an independent, qualified examiner reports on the project's accident potential and safety performance (AUSTROADS 1994, 2009).

RSAs involve applying procedures for looking at the physical elements of the road. The safety problems highlighted are invariably physical in nature. RSAs take principles developed through accident remedial programs that have been found to be effective and apply them proactively (prevention is better than cure).

RSAs provide a number of general benefits, which include reducing

- The likelihood of crashes
- The severity of crashes
- The total costs of crashes to the community

RSAs should be applied at all stages of a road's life: feasibility, draft and detailed design, pre-opening, and existing roads. The preference is for safety audits to be conducted at the earlier stages of the road design process. RSAs are conducted by an experienced and trained person or team (preferred) but independent from the designer.

RSAs identify who the users of a road are or will be and consider the users' ability to deal with traffic and traffic devices, especially the aged, children, and cyclists. RSAs are conducted using checklists, judgment, and knowledge (AUSTROADS 1994).

RSAs incorporate the concept of quality assurance, a management process in which the provider of goods or services assures the customer or client of the quality of those goods or services, without the customer or client having to check each time. A set of procedures is designed and implemented to ensure that agreed standards are met. This involves getting it right the first time!

RSAs can raise a number of important legal issues, such as negligence. Any person injured as a result of the alleged negligence of a highway authority must prove, upon the balance of probabilities, that the authority owed a duty of care, failed to act reasonably, and caused the damage suffered.

Road designers and managers have a duty of care but may have qualified immunity as public authorities. However, failure to take reasonable care by rejecting or not implementing audit recommendations may allow litigation to proceed.

A review of a number of evaluations of RSAs concluded that they are an important tool for minimizing road trauma (AUSTROADS 2002).

6.3.2 Steps in Conducting a Road Safety Audit

The following steps are recommended when conducting a road safety audit:

a. Auditor selection
   - Independent, trained, skilled, and right aptitude
b. Background information
   - Necessary and relevant information
   - Project intent, site data, plans, and drawings
c. Commencement meeting
   - Background information
d. Documents
   - Review documents
   - Form conclusions
e. Site inspection
   - Consider perspective of all road users
   - Photographs and videotapes
f. Report
   - Hazard identification
   - Recommendation about corrective actions
   - Nature and direction of solution
   - Findings
   - Formal statement signed and dated by auditor
g. Completion meeting
   - Auditor and client
   - Familiarization/clarification
h. Following up
   - Implementation decision
   - Acceptance/rejection

Safety does not come about automatically by complying with standards and guidelines, but they are a good starting point in any design.

6.3.3 Safer Road Design

A number of important questions relating to the design of a road should be considered, such as:

- Can the design be misunderstood by the road users?
- Can it cause confusion?
- Does it create ambiguity?
- Does it provide insufficient information?
- Does it provide too much information?
- Does it provide inadequate visibility or obstructions to vision?
- Does it contain obstacles or "booby traps"?

If the answer is yes to any of these questions, subsequent open questions need to be asked, such as how, why, when, and where.

6.3.3.1 Safe Road Environments

There are a number of essential elements in safe road environments that

- WARN the driver of any substandard or unusual features.
- INFORM the driver of conditions to be encountered.
- GUIDE the driver through unusual sections.
- CONTROL the driver's passage through conflict points or sections.
- FORGIVE the driver's errant or inappropriate behavior.

A safe road environment is one that provides no surprises, with a controlled release of information.

6.3.3.2 Safe Road Design Principles

A number of road design principles are important for road safety, such as:

- Considering all road users (including pedestrians and cyclists)
- Determining a design speed that is compatible with project objectives and appropriate for the adjoining roadside activity and terrain
- Considering the design context, especially the road hierarchy
- Coordinating the vertical and horizontal alignments
- Having simple layout and control at intersections
- Ensuring adequate visibility at intersections

6.4 OUTLINES OF TRAFFIC SAFETY ISSUES

6.4.1 Trends and Causes of Accidents

It is clear that the lack of improvement in traffic safety is one of the biggest issues from which we are suffering all over the world. Due to the recent intensive progress in economies and industries in most Asian countries, rapid motorization can be observed and thus the number of traffic accidents tends to increase year by year. Accordingly, it is very important for us to propose effective countermeasures to mitigate traffic accidents and to do the related basic research, especially on both mechanism and impact of traffic accidents.

Although it seems that Japan is facing a relatively mature era in terms of motorization compared with the other Asian countries, there is still an increasing trend of the number of accidents and persons injured as shown in Figure 6.3.

Figure 6.4 shows the major causes of traffic accidents reported by the national police agency in Japan. As shown in the figure, most

Number of:	1995	2006
Fatalities	10,679	6,352
Injured persons	922,677	1,098,199
Accidents recorded	761,789	886,864
Vehicle registered	70,100,000	79,452,557
License holder	68,563,830	78,798,821

FIGURE 6.3 Statistics related to traffic accidents in Japan (1995 versus. 2006).

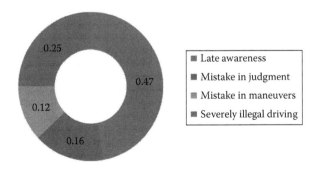

FIGURE 6.4 Major causes of accident in Japan.

accidents might have been caused by human error, including late awareness of dangerous events in traffic and mistakes in both judgment and maneuvers. Accordingly, it can be said that it is necessary for us to enhance the basic research on the causes and mechanism of accidents in order to make countermeasures for traffic safety more effective.

In addition, we have to pay attention to the high percentage of elderly persons (above sixty-five years) and young generations (below twenty-four years) among fatalities in Japan.

Considering the serious situation in terms of traffic safety in Japan, the prime minister provided us with the following statement in January 2003: "We are aimed at achieving the safest roadways all over the world and making the number of fatalities less than 5,000 per year during the next ten years." Based on this statement, the Japanese government has initiated various countermeasures to improve traffic safety, such as improvement in traffic regulation and strict enforcement, improvement in traffic safety devices, enhancement of passive safety of vehicles, and so on.

6.4.2 Difficulty of Traffic Safety Studies

The ultimate goal of traffic and transportation engineering is to make traffic flow on road networks more efficient, smoother, and safer. Compared with traffic safety aspects, much research has been carried out for evaluating the level of service provided by a road transportation system from the viewpoints of efficiency and smoothness of traffic flow, and several indices are available—for example, travel speed, jam length, maximum throughput of a network, and so on. In past studies, safety has been frequently discussed on the basis of the change in the number of accidents recorded. However, it seems to be difficult to evaluate the safety aspect of traffic flow based only on the data of recorded traffic accidents in an objective or statistical manner, because accidents are a rare event when we focus on a certain road section.

As mentioned in the previous subsection, it can be said that human error might be a predominant cause of accidents. (For example, in Japan, around 75% of accidents are caused mainly by human error.) Accordingly, it is expected that the enhancement in basic research on both causes and mechanisms of traffic accidents might contribute to improvement in countermeasures for traffic safety. However, there have been difficulties in collecting suitable data for analysis of mechanisms of traffic accidents. This might be a reason why less accumulation of safety research can be found compared with that of traffic flow analysis. Recently, as mentioned in the next subsection, rapid progress in intelligent transportation systems (ITS) has led to a drastic change in data collection, and it has become possible for us to collect data directly on the occurrence of accidents and to collect data on interaction between vehicles and corresponding conflicts. The innovative changes in data collection will bring about better understanding of the mechanisms of traffic accidents through accumulating the basic research.

6.5 ITS FOR IMPROVEMENT IN TRAFFIC SAFETY

6.5.1 Important Viewpoint for Countermeasures for Improving Traffic Safety

Improving traffic safety is an urgent and important issue to be mitigated as soon as possible. As mentioned, it is clear that human error is a critical factor that might cause traffic accidents. In addition, there are various factors that might exert a certain influence on occurrence of traffic accidents. Traffic conditions and driving environments, such as weather conditions, physical structure, and geometry of roadways, should be included in the factors considered. Considering the complexity in the mechanisms of traffic accidents, it might be true that just one or two countermeasures are not enough for drastically improving traffic safety.

Here are three "*Es*" representing the important aspects to be considered for developing countermeasures for improving traffic safety:

a. Education
b. Enforcement
c. Engineering

Road transportation is one of the important social systems in enhancing socioeconomic activities. In other words, everybody belonging to our modern society is required to participate in the road transportation system as its users. In this sense, enhancement in education related to traffic safety is very important in terms of providing people with both basic knowledge about traffic safety issues and opportunity to review their behavior from the viewpoints of traffic safety in the long run. But actually there seems to be a lot of room for improvement in terms of traffic safety education at the elementary or junior high school level in Japan. No systematic education has been provided to students considering the educational guideline issued by the Ministry. This is one of major issues to be discussed intensively.

Enforcement is expected to play an important role in directly reducing the effects of accidents. The following are examples of recent changes in traffic regulation aimed at improving traffic safety in Japan:

a. 1996: Compulsory usage of child safety seat
b. 2002: Strict enforcement of regulations for drunk driving with severer fines and penalties
c. 2004: Strict enforcement of regulation for mobile phone usage

Especially, it seems that the change in regulations in drunk driving and their strict enforcement might have contributed to reduction in the number of accidents caused mainly by drunk driving, as shown in Figure 6.3.

6.5.2 Possible Contribution of ITS to Improving Traffic Safety

From the viewpoint of the contribution of engineering to mitigating traffic accidents, ITS are expected to play important roles. The expected roles of ITS may be mainly classified into two categories:

a. Driver assistance to reduce human error leading to accidents
b. Advanced sensing and detection of driver's and/or vehicular behavior

The systems and services included in the first category have been well known as the typical application of ITS for traffic safety. As mentioned, it is reported that around 75% of accidents have been caused mainly by human error in driving. Accordingly, these kinds of systems, such as advanced cruise-assist highway systems (AHS) in Japan, have been developed actively in many countries.

The second expected role of ITS might be technically supported by the recent progress in image processing, probe vehicle, drive recorder, and so on. As mentioned before, this kind of role might bring about innovative change in data collection and thus we will have better understanding of the mechanism of traffic accidents through accumulating basic research. The better understanding of causes and mechanisms of accidents is expected to bring about a higher possibility to apply more suitable countermeasures to improve traffic safety.

The key technological element is the vehicle–highway cooperation supported by sensing and communication technologies. For example, it is expected that 5.8 GHz of DSRC (dedicated short-range communication) may make high-speed communication of large amounts of information between vehicles and highways possible. Also, the recent progress in sensing technologies enables the vehicle itself to detect obstacles on the roadway automatically. For example, a radar system using an ultra red beam can detect an obstacle on a roadway in foggy conditions, and thus some service provided by ITS may help information collection and judgment of drivers.

6.6 A REVIEW OF TRAFFIC CONFLICT ANALYSES

In previous research, a "traffic conflict index" has often been used for evaluating the safety of a road section. A traffic conflict is defined as "an observable situation in which two or more road users approach each other in space and time to such an extent that there is a risk of collision, if their movements remain unchanged." Because the frequency of observed traffic conflicts is generally high compared to accidents, the traffic conflict index might be a suitable one. There are many studies evaluating traffic conflicts subjectively by the observer at the initial stage of the research. One of the earliest studies is the development of the so-called traffic conflict technique (TCT) by Perkins and Harris (1967). TCT is useful for evaluating the traffic conditions that might cause a dangerous situation based on recorded near-accidents. Perkins and Harris proposed to record the observable activities of drivers to avoid an accident in the case of a traffic conflict. Also, Spicer (1971), Cooper (1977), and Zimmerman, Zimolong, and Erke (1977) studied the subjective evaluation of traffic conflicts. All these studies have in common that the traffic conflicts are recorded by the observers based on their own criteria established in advance. Therefore, these studies might be criticized for diversity in the judgment on whether there is a traffic conflict or not.

Both TTC (time to collision, Hayward 1972) and PET (postencroachment time; Allen, Shin, and Cooper 1978) are indices to evaluate the conflict objectively and are widely used in traffic safety analysis. Computing these indices requires microscopic data on vehicular movements, but in the 1970s it was difficult to obtain these kinds of data, due to the huge amount of effort and time necessary. The recent progress in image processing techniques enables us to obtain the microscopic data on vehicular movements easily in time and space from video image recording of traffic flow dynamics. Accordingly, applying the indices derived from a traffic conflict index is now a useful and feasible methodology in traffic safety analysis.

TTC is one of the indices widely used to evaluate traffic conflict. In order to estimate TTC, it is assumed that the vehicles keep the same speeds and do not change directions from the point in time when TTC is required to be estimated. Based on this assumption, the remaining time until the following vehicle collides with the tail of the vehicle in front is estimated as TTC. Though both the concept and the calculation method of TTC are very clear and easy to understand, TTC has a drawback in that its finite value cannot be estimated if the leading vehicle travels faster than its following vehicle. However, there may also be a danger of vehicle collision in the case that the leading vehicle travels slightly faster than its following vehicle, but the distance between these two vehicles is very short. Especially, if the leading vehicle applies its emergency brake for some reason, there is a large possibility that the following vehicle cannot stop safely in time and therefore will collide with the leading vehicle. TTC does not consider this kind of dangerous situation.

Based on the preceding discussion, we propose a new index called PICUD (possibility index for collision with urgent deceleration) to evaluate traffic conflict. It evaluates the possibility that two consecutive vehicles collide if the leading vehicle applies its emergency brake. PICUD is defined as the distance between the two vehicles at the time when they completely stop. Figure 6.5 and Equation (6.1) outline the way to compute PICUD:

$$PICUD = \frac{V_1^2}{-2a} + s_0 - \left(V_2 \Delta t + \frac{V_2^2}{-2a} \right) \qquad (6.1)$$

where

V_1 and V_2 represent the speeds of the leading and following vehicles, respectively, when the leading vehicle applies an emergency brake

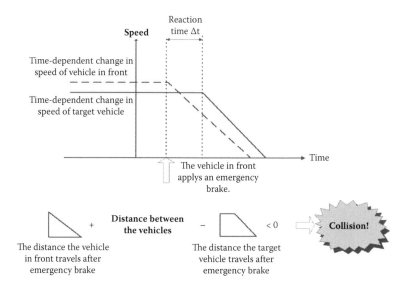

FIGURE 6.5 Computation method of PICUD.

s_0 indicates the distance between the two vehicles at the point in time when the leading vehicle applies the emergency brake

Δt indicates the reaction time of the driver of the following vehicle

a represents the deceleration rate of both the leading and the following vehicle

Here, a is assumed as 3.3 m/s^2 and the reaction time Δt is assumed as 1.5 s. A negative value of PICUD means that there is a possibility that a sudden braking maneuver of the leading vehicle might cause a collision even if the driver of the following vehicle also applies the emergency brake.

6.7 UTILIZATION OF VIDEO IMAGE DATA FOR CONFLICT ANALYSIS

6.7.1 Outlines

In order to make the countermeasures for improving traffic safety more effective, we should understand the mechanism of traffic accidents based on the research of traffic accidents in a scientific and quantitative manner. To do so, it is necessary for us to collect the data on drivers' behavior and interaction among vehicles, motorcycles, bicycles, and pedestrians. To simplify our discussion here, we focus on the interaction among vehicles.

It is possible for us to apply several approaches for investigating both drivers' behavior and interaction among vehicles, such as field experiments using test vehicles in which video cameras and sensors are installed, in-laboratory experiments using driving simulators, and application of video image data to extract vehicle trajectory. Compared with the other approaches, the application of video image data has the following strong points:

a. It is possible to obtain data directly on behaviors of vehicles and their interactions in the real traffic environment.
b. It is possible to investigate diversity in vehicular behaviors.
c. The quantitative data on vehicle behavior represented by vehicle trajectory can match easily with the traffic phenomenon recorded by video image.

Of course, the distance covered by a video camera is limited, so it is necessary to develop a processing technique both to connect the vehicular trajectories obtained by different cameras and to smooth the connected trajectories, in order to obtain the vehicular data along the analyzed section with a certain distance, such as 1 km and so on. In this section, an example of conflict analysis is introduced to show how we can actually apply video image data of traffic flow for analyzing conflicts between vehicles (Uno et al. 2006).

6.7.2 An Example of Conflict Analysis Using Video Image Data

6.7.2.1 Outline of Studied Section

The road section analyzed is one of the steep curve sections of the Meihan National Highway—a 31.6 km long, access-controlled trunk road that connects Tenri of Nara Prefecture with Kameyama of Mie Prefecture, Japan. Meihan National Highway started its operation in 1965 and completed its widening to a four-lane highway in 1977. The speed limit of the road is 60 km/h, and motorcycles with an engine of less than 125 cc as well as bicycles and pedestrians are not allowed to use it. The highway connects Higashi (Eastern) Meihan Motorway and Nishi (Western) Meihan Motorway—a major auto route between the Kansai metropolitan area (including Osaka, Kyoto, and Kobe) and the Chubu metropolitan area (including Nagoya).

On the Meihan National Highway many accidents have been recorded, and therefore serious safety measures should be applied to this road as soon as possible. Reasons for the high number of accidents are as follows:

a. Due to the intensive motorization of the society in Japan, a huge number of trucks travel along the Meihan National Highway.
b. A lot of vehicles are traveling along the highway above its speed limit (60 km/h) because it connects the two motorways mentioned earlier, which have a speed limit of 80 km/h.
c. Meihan National Highway includes a lot of steep slopes and sharp curves.

The road section where the highest numbers of accidents are observed is the "omega-shaped curve" between the Tenri-higashi Interchange and Fukuzumi Interchange. Especially along the westbound curve, a high number of accidents are observed. The curve includes several steep down slopes with a gradient of 5% to 6% and sharp curvatures with a diameter of less than 300 m. We focus on the vehicular movement at the Nakahata Curve that is part of the omega-shaped curve. Because there is a short, straight section between two consecutive sharp curves, traveling along Nakahata Curve might require complicated maneuvers from the drivers so as not to lose control of their vehicles.

6.7.2.2 Traffic Accidents Occurred at Nakahata Curve

Figure 6.6 indicates the location and the types of accidents recorded at Nakahata Curve in 2002. For convenience and considering the horizontal alignment of the road, the studied section is assumed to be classified into the nine subsections shown in Figure 6.6.

Judging from Figure 6.6, it can be said that the section with the highest number of accidents is the third transition curve between the short, straight section and the second curve, which has a diameter of 300 m. The number of accidents in this section reached twenty-four in 2002. Further, the second highest number of accidents is recorded in the first curve, which has a diameter of 200 m.

In Figure 6.6, the accidents that occurred along this section can be classified into the following types: (a) lane departure and collision with median barrier, (b) other accidents involving only a single vehicle, (c) rear-end collisions of two or more vehicles, and (d) other accidents involving multiple vehicles. Judging from Figure 6.6, lane departure and subsequent collision with the median barrier is predominant compared with other types of accidents at the third transition curve with the highest number of accidents recorded.

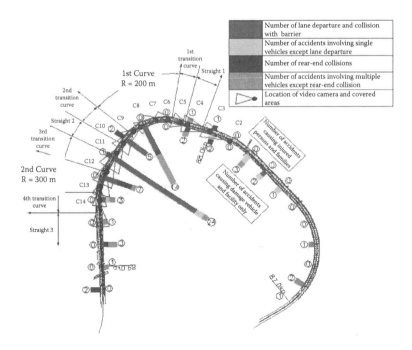

FIGURE 6.6 (See color insert.) Location and type of accidents recorded at Nakahata Curve.

6.7.2.3 Data Used for Analysis

Along Nakahata Curve, thirteen digital video cameras are installed to observe almost 900 m of the road section and to obtain the video images of traffic flow. Here, the image processing technique including background subtraction is applied to extract the vehicle trajectory data (VTD) from the video image. The distance of the road section covered by each video camera depends upon both the height of the camera installed and the road alignment, and it varies from 70 to 100 m. Figure 6.7 indicates the still images obtained from the cameras mentioned earlier.

For each pair of two consecutive cameras, the area covered by the downstream camera overlaps with the one covered by the upstream camera in order to enable us to trace the vehicle movement continuously throughout the studied section. The length of the overlapping area is about 10 m for each pair. The video data used were collected from October 16 to November 3, 2003, and include the data on traffic flow under both sunny and rainy weather conditions.

Here, the VTD are assumed to be composed of a series of dots representing the location of vehicles on the x–y axis. Figure 6.8 shows the concept of an x–y axis setting. It is assumed that the y-axis corresponds with the direction of carriageway and the x-axis is orthogonal to the y-axis at every point. Figure 6.9 shows examples of raw VTD obtained

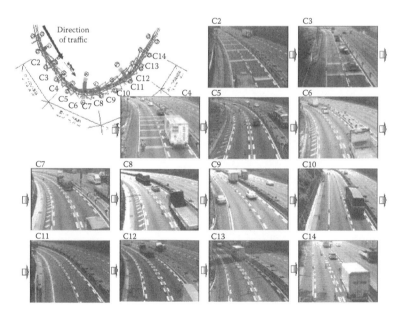

FIGURE 6.7 (See color insert.) Examples of images obtained from the video cameras.

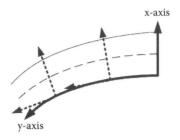

FIGURE 6.8 Concept of x–y axis setting.

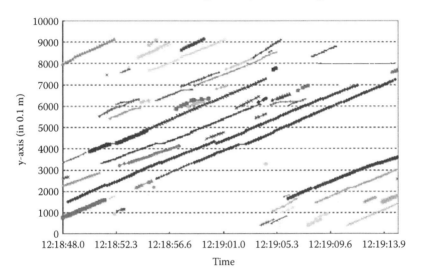

FIGURE 6.9 (See color insert.) Examples of raw vehicle trajectory data.

from the video image. It shows the vehicle position on the y-axis against time. After eliminating the error data, the vehicle trajectory data provide us with data on the vehicle position, its speed, its spacing, and the relative speed between two consecutive vehicles. Concerning the accuracy of the data, although it is not reasonable to expect that these data are 100% accurate due to the influence of measurement errors, it is expected that they are suitable for calculating the indices to grasp roughly the spatial and temporal change in the potential danger of traffic accidents occurring along the studied road section.

6.7.2.4 An Example of Evaluating the Potential Danger of Lane Departure

Focusing on the types of accidents in the studied section, the lane departure and collision with median barrier are predominant, as mentioned before. Accordingly, we propose *the estimated centrifugal acceleration of vehicles* as an index to evaluate the potential danger of lane departure and collision with the median barrier.

Here, the centrifugal acceleration of the vehicle is estimated in order to evaluate the stability of the vehicle movement at the curved section considering the vehicle dynamics:

$$[Estimated\ centrifugal\ acceleration] = V^2/R \qquad (6.2)$$

In this equation, V represents the speed of the vehicle obtained from its trajectory data, and R represents the diameter of the curve. If a vehicle runs at very high speed along the steep curve, the large centrifugal acceleration might affect the vehicular movement and the driver might lose control in the worst case. The inequality (Equation 6.3) represents the condition under which the vehicle running along the curved section is not likely to be out of control:

$$g(i+f) \geq \frac{V^2}{R} \qquad (6.3)$$

where g represents the gravitational acceleration, i represents the super-elevation, and f represents the coefficient of friction between tire and road surface. If a vehicle runs at very high speed along the curve with small diameter and the combination of V and R does not satisfy the inequality (Equation 6.3), there is a high possibility that lane departure and a collision with the median barrier might occur.

6.7.2.5 Analysis Based on Estimated Centrifugal Acceleration

As explained previously, there are many observations on lane departure and collision with the median barrier under rainy conditions in the studied section. In this subsection, whether or not the vehicle is likely to slip with the consequence of lane departure and a collision with the median barrier is analyzed, based on the estimated centrifugal acceleration. Figure 6.10 shows the average and standard deviation of the centrifugal acceleration under rainy conditions. Because of the smaller diameter of the first curve (200 m) compared with that of the second curve (300 m), the estimated centrifugal acceleration of the first curve tends to become larger than that of the second curve. Further, due to the higher speed of vehicles driving in the second lane compared with that of those in the first lane, the estimated centrifugal acceleration of the second lane tends to become larger than that of the first lane. Clearly, the largest centrifugal acceleration is observed at the first curve, and it seems that the vehicles in this curve are strongly affected. In order to evaluate

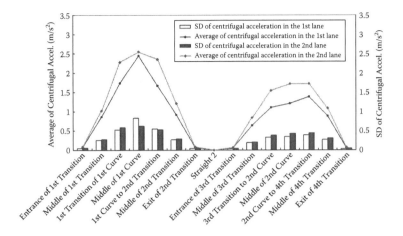

FIGURE 6.10 Average and standard deviation of centrifugal acceleration under rainy condition.

the possibility of the slip of vehicle, it is useful to estimate the critical speed and centrifugal acceleration to satisfy the following relation:

$$g(i+f) = \frac{V^2}{R} \qquad (6.4)$$

Equation (6.4) can be derived by replacing the equal sign with the inequality in Equation (6.3). Concerning the design parameters of the first curve, the diameter is 200 m and the super elevation is 0.0826. Following previous research about the coefficient of friction between tire and road surface, it is assumed to be 0.3 in rainy conditions. Under these assumptions, the critical speed and centrifugal acceleration estimated become 98.6 km/h and 3.75 m/s² in rainy conditions. Considering both average and standard deviation of the estimated centrifugal acceleration shown in Figure 6.10, it can be said that at least several percent of the vehicles must be running along the first curve at a speed higher than the critical one, and that these vehicles are seen as unsafe regarding the possibility of lane departure and a collision with the median barrier.

6.8 STRUCTURAL SAFETY

This section discusses the structural safety of bridge structures against severe environments, especially extraordinary traffic load. The bridges in Asian countries are often subjected to extraordinary loads of overloaded vehicles, which accelerate the fatigue damage of the bridges. To assess the safety and also adjust the design code of the bridges in Asian countries, traffic information on site should be known: Bridge weigh-in-motion (BWIM) is one of the simple methods to assess the traffic, especially applying loads from the passing vehicles, just by using the responses of the bridge. Thus, herein the fundamental instruction of BWIM is described as well as a case study to assess the traffic load in Bangkok, Thailand.

6.8.1 Overloaded Vehicles in Asian Countries

In recent years, many countries in Asia have been growing rapidly and expanding their economic possibilities. Associated with this growth, infrastructures such as bridges, tunnels, and highways have been needed and built as fast as possible, and thus the requirements for these infrastructures have sometimes been beyond their performance because the actual environment is likely to exceed the design assumptions. In reality, unexpected overloads on the bridges and roads transcend their limitation because the trucks carry as much as possible (sometimes almost impossible amounts) to increase transport efficiency. Such overloaded vehicles are often found in Asian countries and are not officially allowed by the regulation of their own countries. In some cases, overloaded vehicles are unofficially permitted by police officers who do not understand their serious threat to structures. Note that overloaded vehicles usually move over the structures at night when police officers are not on duty.

6.8.2 Impacts of Heavy Vehicles on Structures

Overloaded vehicles will cause serious damage to infrastructures, especially road pavements and bridges. Rarely, the structures collapse by direct action from overloaded vehicles because, in most cases, the bearing capacity of bridges is much higher than the applied load. Figure 6.11 shows one example of a bridge collapse in Cambodia. In this case, the capacity of a temporary bridge (Bailey Bridge) was not known, and the load of a heavy vehicle transcended its ability. However, most of the time, the bridges may fall down due to fatigue damage, not by direct action of load. Figure 6.12 shows the fatigue cracks of concrete slab of a road bridge (left) and crack of a cross beam (right) in Japan. These damages may lead to fatal collapse of a bridge. Structures bearing repeated loads are always subject to danger of fatigue failure.

Fatigue is a kind of cumulative damage, and a material subject to repeated loads fails when the damage accumulates to reach its limitation. Among the several theories to estimate fatigue damage, Miner's law is the most useful so far. In this theory, the cumulative damage shows

FIGURE 6.11 Bridge collapse in Cambodia (Bailey Bridge).

FIGURE 6.12 (See color insert.) Fatigue cracks of a concrete slab (left) and cross beam (right) in Japan.

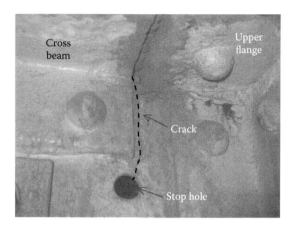

FIGURE 6.12 (*Continued*)

a linear relationship with the number of cycles in a logarithmic axis assuming that the damage may accumulate by the amplitude of stress. For instance, the fatigue damage of steel material increases in proportion to the third power of the number of cycles. This means that double heavy trucks give eight times larger damage to the bridges. Therefore, the impact of overloaded vehicles on bridge structures cannot be ignored since they significantly decrease the safety of structures and should be evaluated on the basis of measured load data.

6.8.3 Bridge Weigh-in-Motion

In Japan, the load limit of heavy vehicles was changed from 20 tons to 25 tons in 1992 and, according to this revision, the design load was also revised to the B-type live load in 1993. Thus, to know the actual live load is still of importance in Japan, especially when the remaining life of a bridge is estimated or the reason of fatigue failure is evaluated. The bridges or roads in Japan as well as in Asian counties have been subjected to larger live loads due to overloaded vehicles than expected, which may accelerate fatigue damage of the structures. For structural safety reasons or administrative regulations, several kinds of monitoring for live loads have been conducted so far. In general, live load monitoring can be categorized into the following two methods:

  a. Direct method: The live load is directly measured by instruments such as truck scales or axle meters.
  b. Indirect method: The live load is estimated using the responses of other structures, such as a bridge.

An axle load scale located in front of the toll gate of a highway is one of the direct measurements. The axle load scale can detect the axles at relatively lower speed (less than 20 km/h) and thus this measurement can be available only in the place where trucks pass at lower speed, such as toll gates. In this system, to avoid the second axle passing on the scale before the first one is out, the sensing part of the scale is very short—a few centimeters—and this leads to unstable measurement due to the vibration of moving vehicles. Thus, in Japan, due to its (in)accuracy, the axle load scale is only used for warnings, not for regulation. When police regulate heavy vehicles, they use truck scales. The truck scale can measure the axle load accurately but only when the vehicle is stopped.

Bridge weigh-in-motion is one of the indirect methods and it uses the response of bridges. This system is very cheap comparing with the other systems. BWIM can be also divided into three categories according to the response members:

a. Strain responses of main girders
b. Strain responses of the second members
c. Responses of cracks on the concrete slabs

Table 6.1 is the summary of BWIM proposed so far by Japanese researchers, according to their principle of the estimation.

6.8.3.1 Estimation by the Strain Responses of Main Girders

BWIM was proposed in the 1970s by Moses (1979) to estimate the fatigue damage of road bridges in the United States. In his theory, the axle loads are estimated by minimizing the error between the assumed influence lines and the obtained responses of main girders using the least squares method. Layout of this estimation is shown in Figure 6.13. Note that this method needs the information of vehicle's velocity and distances of axle loads, as well as influence line, in advance. To obtain this information, some researchers use optical switches, but in Japan the responses of concrete slab are mainly used because they respond very acutely to each axle.

When the influence line is fitted by a polynomial, the influence function by one axle is assumed as follows:

$$f(x) = a_0 + a_1 x + \cdots + a_n x^n \tag{6.5}$$

Now, for instance, the influence line of the vehicle with two axles, P_1 and P_2, at the interval of L can be given by

$$h(x) = P_1 f(x) + P_2 f(x - L) \tag{6.6}$$

TABLE 6.1 BWIM

Type	Principle	Sensing Members	Notes
I	Influence lines of main girders are used. The influence line is interpolated by a polynomial or just a linear function, and then the axle load is estimated by minimizing the error of the function and obtained data.	Lower flange of main girders (influence line) Optical switch (velocity and wheel base)	Velocity of passing vehicles must be used. The accuracy decreases when several vehicles are passing on the same bridge at once.
II	Influence line of stringer The area of responses in time is proportional to gross weight.	Stringers (influence line, velocity)	Vehicle location must be known and the estimation cannot be conducted when several vehicles are on the bridge.
III	Influence line of support reaction The gap in the response corresponds to the axle.	Vertical stiffeners (influence)	No velocity information is needed. The structures without vertical stiffeners cannot be applied.
IV	The axle is estimated by the relationship between crack responses of RC slabs and the axles.	Concrete slab (response of cracks)	No velocity information is needed. The accuracy increases as the installed sensors increase.
V	The influence line of main girder is used but fitted by linear function. Similar method to type I.	Main girder (influence) and concrete slab (velocity, lane and wheel base)	Velocity of passing vehicles must be used. The accuracy decreases when several vehicles pass on the same bridge at once.

Then the vehicle with n axles causes the response of the main girder in the form of

$$M(t_k) = \sum_{i=1}^{n} P_i f_i(t_k) \tag{6.7}$$

where f_i is the value of the influence line where P_i is located at t_k. This can be obtained by the wheel base L and velocity. Assuming the measured

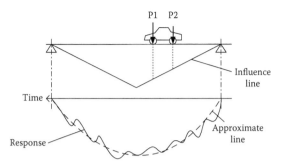

FIGURE 6.13 BWIM based on the response of the main girder.

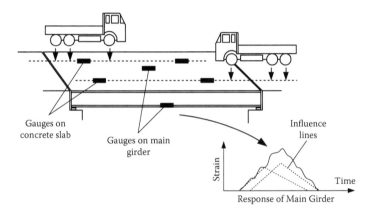

FIGURE 6.14 Layout of BWIM system.

value is M^*, then the evaluation function can be given by the following with the measurement time t_1 to T (increment is t_k):

$$J = \sum_{k=1}^{T} \{M(t_k) - M^*(t_k)\}^2 \qquad (6.8)$$

Then, each axle P_i can be obtained by minimizing the preceding evaluation function.

In the past, research based on the preceding theory, the number of axles, and the wheel base was obtained by optical switch. The other system based on the response of main girders also uses that of concrete slab for axle recognition and wheel base estimation (Tamakoshi et al. 2004). In this system, total weight of the vehicle is estimated by the response of the main girder, and the axles are also distinguished by the response of concrete slab, as shown in Figure 6.14. The response of concrete slab

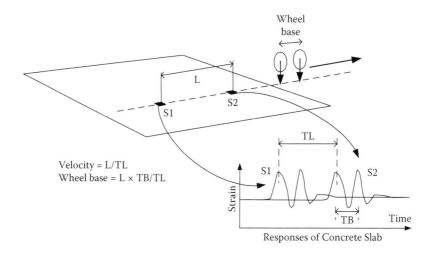

FIGURE 6.15 Recognition of velocity and wheel base of a passing vehicle.

is more sensitive than that of the main girder so that each axle can be distinguished. As can be seen in Figure 6.15, two sensors on the slab are installed away from each other in one stream line. Thus, the wheel base (distance of axles) and velocity can be recognized by using the response lag between these two sensors because each peak corresponds to an axle load.

First of all, the wheel base and the number of axles of a vehicle are obtained by the two sensors on the slab (S1 and S2 in Figure 6.15). Then, assuming that the monitored span is a simple beam, the strain response of the main girder due to an axial load P_i can be given by

$$\varepsilon_i(x) = EZ \frac{x'}{2} P_i \qquad (6.9)$$

where Z is section modulus and E is elastic modulus x', also defined as

$$x' = \begin{cases} x & (0 \leq x \leq l/2) \\ l-x & (l/2 \leq x \leq l) \end{cases} \qquad (6.10)$$

where l is a span length (Figure 6.16). In this system, EZ should be obtained by calibration using the truck whose weight is known by the other system.

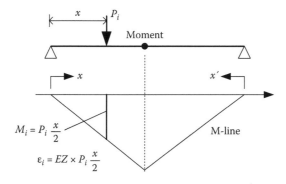

FIGURE 6.16 Strain response of main girder.

Then, the total response by a vehicle with n axles can be expressed as

$$\varepsilon(x) = \sum_{i=1}^{n} \varepsilon_i(x) = EZ \sum_{i=1}^{n} \frac{x'}{2} P_i \quad (6.11)$$

Note that the number and position of P are already known by the sensors on the slab. Thus, we can estimate each P by minimizing the errors between measured and calculated values using the least squares method as follows:

$$J = \sum_{j=1}^{m} \{\varepsilon(x_j) - e(x_j)\}^2 = \sum_{j=1}^{m} \left\{ \sum_{i=1}^{n} \varepsilon_i(x_j) - e(x_j) \right\}^2 \to \min \quad (6.12)$$

where $e(x_j)$ is measured strain and x_j denotes the jth time period.

6.8.3.2 Estimation by Responses of the Second Members

Live loads are also estimated by the responses of the second members such as vertical members in truss structures or stringers in girder bridges (Ojio et al. 2001, 2003). These methods are based on the fact that the area of responses is proportional to the gross weight. In these methods, the response signal in time space is used and the response of the bridge $g(x)$ is assumed to be given by the following convolution of the influence:

$$g(x) = \sum_{k=1}^{n} P_k f(x - L_k) \quad (6.13)$$

where x is the position of the vehicle, P_k is the kth load, is the influence function and L_k is the distance between the first axles to the kth axle. Then the area can be given by the following:

$$A = \int_{-\infty}^{\infty} g(x)dx = \sum_{k=1}^{n} P_k \cdot \int_{-\infty}^{\infty} f(x)dx = W \cdot \int_{-\infty}^{\infty} f(x)dx \qquad (6.14)$$

where W is gross weight of the vehicle and is constant to all kinds of vehicles. Then, based on the vehicle weight W_c and the response area of the well-known gross weight A_c, gross weight of any vehicle can be detected by this method:

$$W = \frac{A \cdot W_c}{A_c} \qquad (6.15)$$

Since the area of influence line is a function of distance, the area can be obtained by the response in time domain by assuming that the velocity is constant. Letting the response of the vehicle with the velocity of v be $r(t)$, the response area can be obtained in the form of

$$A = \int_{-\infty}^{\infty} g(x)dx = v \cdot \int_{-\infty}^{\infty} r(t)dt \qquad (6.16)$$

Another method has been proposed using the influence line of support reaction, which can be estimated by the strain response of vertical stiffeners. In this theory, axle loads are estimated by the sudden changes in the response caused by entry of the vehicle, and wheel bases are also obtained by the time difference between the rise and last of the response, which corresponds to the time of entrance and exit. As shown in Figure 6.17, the sudden change in the response is proportional to the corresponding axle load.

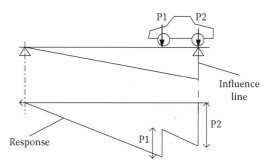

FIGURE 6.17 BWIM using the influence of reaction force.

6.8.3.3 Estimation by Crack Response of Reinforced Concrete Slab

Crack width in the main reinforcement direction of reinforced concrete (RC) slab corresponds to each axle. Matsui and El-Hakim (1989) proposed the estimation theory based on the response of the cracks. However, because the cracks are highly sensitive to the alignment of the axle, many cracks and sensors are required to catch the axle under the sensors.

6.8.3.4 Error Factors of BWIM

The accuracy of BWIM slightly depends on the running condition of passing vehicles. The accuracy of estimated axle load, gross weight, and wheel base is different in different conditions.

First of all, simultaneous loading such as parallel running and continuous running is the largest factor to decrease the accuracy. The probability of simultaneous loading increases in the case where a bridge has many girders and lanes and the case where a bridge has continuous girders, since the influence line of each member becomes longer. The situation of simultaneous loading is shown in Figure 6.18.

The main issue associated with the use of main girders is the assumption of constant velocity. Thus, when the velocity is changed in the sensing section (for instance, sudden acceleration, braking, or traffic jams), it causes much error in the estimation. However, the difference of constant velocity does not have much influence on the estimation when it is constant. It is reported that a BWIM system using main members can estimate the gross weight with accuracy of 5% when a single vehicle passes on the bridge. But in the case of parallel running, the accuracy of estimated gross weight decreases to 50% and that of axle load decreases to 70%. In the case of continuous running, the accuracy decreases to less than 20% in the gross weight but 40% in the axle load.

To avoid simultaneous loading, the estimation can be done by second members that have shorter influence lines. For instance, in the case of steel slab, the stain of ribs responds to the axle exactly on the lane above the rib and the use of the rib can reduce the influence of simultaneous

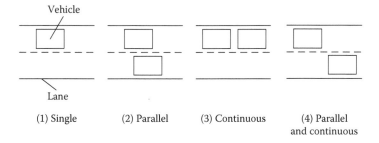

FIGURE 6.18 Simultaneous loading.

loading. Another method focuses on the stringer that is only sensitive to the axle load on the stringer. These methods have the merit of avoiding the simultaneous loading but the response decreases greatly as the alignment of axles is out of the members.

The method using vertical stiffeners has merit in that the estimation does not require the velocity. This method has high accuracy, such as less than 20% for axle load, even in the case of simultaneous loading. However, this method cannot be applied to the bridge without vertical stiffeners and the accuracy decreases as the velocity increases due to the dynamic effect.

Other factors that decrease the accuracy are vibration of the vehicle and the stiffness of pavement, which is very sensitive to temperature. The vibration of the vehicle directly changes the axle load and the shorter sensing span has more influence on the vibration. It is reported that an error of 20% to 40% occurs in dynamic measurement of the static WIM. The temperature also has much influence on the estimation since the stiffness of pavement is drastically changed based on temperature. Thus, calibration must be conducted at each temperature when the temperature changes greatly during the monitoring period, such as summer and winter.

In Europe, much research has been done on WIM and a code has been proposed recently (COST232 1999). In this code, the location of site, the algorithm of estimation, the required accuracy and the calibration are prescribed.

6.8.4 Case Study in Bangkok, Thailand

Traffic monitoring by BWIM was conducted in Bangkok, Thailand, to observe an actual traffic situation in this city and also to build a typical model of loading in Bangkok. The monitored bridge for BWIM is one of the flyovers crossing the main roads, which are located in the center area of Bangkok. As shown in Figure 6.19, the monitored section has

FIGURE 6.19 Monitored bridge: two independent superstructures stay on common piers.

two superstructures which are independent of each other, on the piers of P3 and P4. The superstructure is a steel concrete composite bridge with two steel main girders and precast concrete slabs. Thus, we installed two independent BWIM systems on them. Layout of the sensor installation is shown in Figure 6.20. In this figure, S1 and S2 denotes the sensor on the slab for axle detection, while M1 and M2 denote the sensor on the bottom flange of main beams to detect the bending moment at the center span. Note that the strain at the center is averaged by two responses obtained from both flanges.

In this monitoring, vehicle type was classified according to the wheel base, and five types of vehicles were assumed according to the detected wheel base in this city. As for two-axle vehicles, the small-sized car indicating a small bus (wheel base is less than 3 m), the middle-sized car indicating a small truck (wheel base is larger than 3 m but less than 4 m), and the large-sized car indicating a heavy truck (wheel base is larger than 4 m) were defined respectively. Note that, basically, the BWIM system can recognize small two-axle vehicles such as passenger cars, but the data of less than 20 kN were ignored in this monitoring because the system is not capable of segregating small vehicles from the noise in the response. In addition to two-axle vehicles, multiaxle vehicles such as three-axle trucks and four-axle trailers were also defined. Although those vehicles can be also classified in more detail by our system, we simply defined one category for each type since it is enough to determine the live load model.

Figure 6.21 shows the relative frequency of gross weight with respect to all kinds of vehicles. Most three-axle vehicles have been detected, but the vehicles with more than three axles were not found by this monitoring. From this figure, it is found that for both exit and entrance lanes the weight distributions of passing vehicles were almost identical

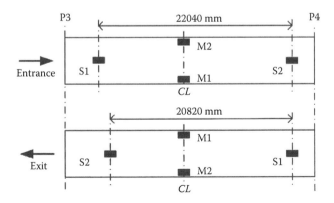

FIGURE 6.20   Location of sensors in the monitored bridge.

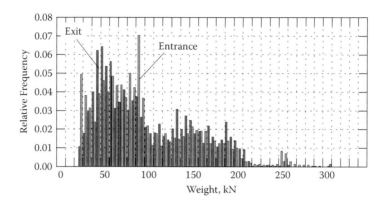

FIGURE 6.21 (See color insert.) Relative frequency of gross weight in exit and entrance lanes.

to each other. Since the estimation accuracy of the exit lane is higher than that of the entrance lane, the detailed analysis was conducted on the entrance lane. However, even for the exit lane, overloaded vehicles were not identified. This may be attributed to the fact that heavy vehicles are banned in this city and also partly because most heavy vehicles tried to evade the monitored bridge in this period and may not have been counted: The bridge was under assessment in addition to the monitoring and the activity of assessment was widely announced to drivers. Thus, the obtained data may represent the situation under severe regulation to control overloaded vehicles because, under the regulation, those vehicles take detours around the city when they pass the city.

Figure 6.22 shows the number of the vehicles passing this monitored bridge for one week in the exit lane. In this figure, the contribution of each type is also illustrated. From this figure, it is found that the large-sized car was dominant in this traffic. Note that the total number of vehicles on day 5 and day 6 is smaller than on the other days because these days were Saturday and Sunday.

Figure 6.23 shows the relative frequencies of four vehicle types. As for the small-sized and middle-sized cars, the distribution shapes are close to log-normal distribution and the averages of gross weight are 46.3 and 72.3 kN, respectively. As for the large-sized car, two main peaks are recognized in the distribution that can be fitted by two different log-normal distributions. For safety reasons, the second distribution with a larger average was adopted as representative here and the average was 164.3 kN. The average of three-axle vehicles was not reliable because the sampling number is too small for these two cases. But in this study we adopted the simple average values to represent this type, and the average values were 178.0 kN.

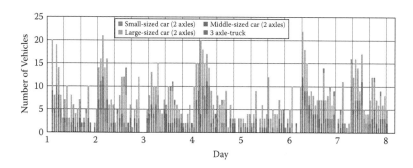

FIGURE 6.22 (See color insert.) Time series of passing vehicles according to the vehicle type (exit lane).

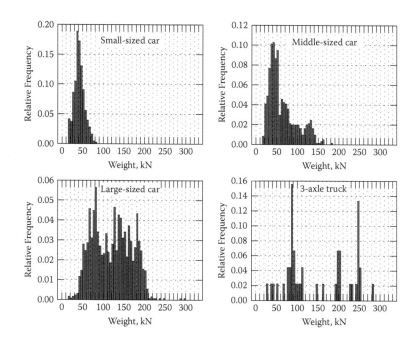

FIGURE 6.23 (See color insert.) Relative frequency of gross weight (according to the types of vehicles).

On the basis of the obtained traffic data, the live load model for a standard bridge was built. The live load model basically complies with the Japanese Specification for Road Bridge (JSRB). Here, a bridge with a 100 m span with two lanes each for upstream and downstream (total of four lanes) is assumed, where the distributed load with the area of 100 m span × 3.5 m width × 2 lanes was applied.

Table 6.2 shows the averaged total weights according to the type of vehicle, based on BWIM. The assumed lengths of the vehicles are also listed in Table 6.3; they were determined by following Japanese standards. As described in the previous section, the BWIM system eliminated the vehicles that were less than 20 kN. Thus, let us assume here that as many passenger cars as other vehicles pass this bridge and they are 13.2 kN and 4.0 m long (half of the traffic may be passenger cars).

Table 6.4 shows the ratio of vehicles based on the measured data as well as the modified ratios in city roads where the heavy vehicle

TABLE 6.2 Measured Weight and Assumed Length of Vehicles

Type	Averaged Total Weight	Assumed Vehicle Length
Passenger car	13.2 kN (assumed)	4.0 m
Small-sized car	46.3 kN	4.0 m
Mid-sized car	72.3 kN	6.2 m
Large-sized car	164.3 kN	8.7 m
Three-axle truck	178.0 kN	10.4 m

TABLE 6.3 Assumed Mix Ratio of Vehicle Types

Type	Measured Ratio	Modified Ratio (City Roads)
Passenger car	50.0% (assumed)	41.9%
Small-sized car	8.8%	7.4%
Mid-sized car	12.8%	10.7%
Large-sized car	27.5%	38.6%
Three-axle truck	1.0%	1.4%

TABLE 6.4 Possible Numbers of Vehicles on the Bridge for Load Design (for Two Lanes)

Type	Number of Vehicles (/two lanes)		Total Weight (kN/two lanes)	
	Normal	Heavy	Normal	Heavy
Passenger car	3.28	9.48	43.3	125.1
Small-sized car	0.58	1.67	26.8	77.5
Mid-sized car	0.84	2.42	60.6	175.0
Large-sized car	3.02	8.73	496.6	1435.0
Three-axle truck	0.11	0.32	19.5	56.4
Total	7.83	22.63	646.9	1869.1
Uniformly distributed load (kN/m^2)			0.92	2.67

(large-sized car and three-axle truck) ratio reaches 40%—standard percentage in Japan. Now we also assume the distance among the vehicles when they are in normal traffic and heavy traffic: 19.4 m for normal traffic and 2.7 m for heavy traffic (those values are assumed using the data of Japanese traffic).

On the basis of the values in Tables 6.2–6.4, the number of vehicles simultaneously on the bridge can be obtained by the following equation:

$$N_i = \frac{L\gamma_i}{\sum_{j=1}^{N_v}\{(L_{Vj} + L_H)\gamma_j\}} \tag{6.17}$$

where
N_i is the number of i-type vehicles by one lane
L is the span of the bridge
γ_i is the mix ratio of i-type vehicles
L_{Vj} is the length of j-type vehicles
L_H is the distance of each vehicle under the assumed mix ratio
N_v is the number of vehicles

The values for the model bridge were finally obtained by multiplying these values by two for two lanes. The calculated values of uniformly distributed load for the model bridge are also listed in Table 6.4.

Comparing with the H20 loading (two-axle trucks) specified in AASHTO, which is 3.0 kN/m² (= 9.2 kN/m/3.048 m), the value even in heavy traffic is less than that of H20. Note that the most serious load assumed in Thailand is HS20 (three-axle trucks) multiplied by 1.3, which is region factor for Thailand; this value (= 5.4 kN/mm²) is much larger than the obtained value. Even though there is heavy traffic in Bangkok and the data were obtained by the regulated traffic, which may not represent a normal situation, the H20 load may be representative for this bridge. However, the monitoring should be continued to obtain a "usual" situation, and the data of other bridges with heavy traffic should be obtained in order to build a suitable live load model in Bangkok.

REFERENCES

Allen, B. L., Shin, B. T., and Cooper, D. J. (1978). Analysis of traffic conflicts and collision. *Transportation Research Record* 667: 67–74.

AUSTROADS (1994). Road safety audit. AUSTROADS publication no. AP-30/94, Sydney.

——— (2002). Evaluation of the proposed actions emanating from road safety audits. AUSTROADS publication no. AP–R209/02, Sydney.

——— (2009). Guide to road safety part 6—Road safety audit. AUSTROADS publication no. AGRS06/09, Sydney.

Cooper, P. (1977). State-of-the-art. Report on traffic conflicts research in Canada. *Proceedings of 1st Workshop on Traffic Conflicts*, Oslo.

COST232 (1999).Weigh-in-motion of road vehicles. Final report.

Hayward, J. C. (1972). Near-miss determination through use of a scale of danger. *Highway Research Record* 384: 24–34.

Lay, M. G. (1985). *Source book for Australian roads*, chap. 28. Vermont, Australia: ARRB Group.

Matsui, S., and El-Hakim, A. (1989). Estimation of axle loads of vehicles by crack opening of RC slab. *Journal of Structural Engineering*, JSCE 35A: 407–418.

Moses, F. (1979). Weigh-in-motion system using instrumented bridges. *Transportation Engineering*, ASCE 105: (TE3): 233–249.

NAASRA, (1988). *Guide to traffic engineering practice—Road crashes*, Part 4. National Association of Australian State Road Authorities, Sydney.

Ogden, K. W. (1996). *Safer roads: A guide to road safety engineering*. Avebury, England: Gower Technical.

Ogden, K. W., and Bennett, D. W. (1989). *Traffic engineering practice*, 4th ed. Clayton, Australia: Department of Civil Engineering, Monash University (chap. 6.7).

Ojio, T., Yamada, K., Kobayashi, N., and Mizuno, Y. (2001). Development of bridge weigh-in-motion system using stringers of steel plate girder bridge. *Journal of Structural Engineering*, JSCE 47A: 1083–1091.

Ojio, T., Yamada, M., Wakao, and Inoda, T. (2003). Bridge weigh-in-motion using reaction force and analysis of traffic load characteristics. *Journal of Structural Engineering*, JSCE 49A: 743–753.

Perkins, S. R., and Harris, J. L. (1967). Traffic conflict characteristics—accident potential at intersections. *Highway Research Record* 225: 35–44.

Ruediger, L., and Mailaender, T. (1999). *Highway design and traffic safety engineering handbook*. New York: McGraw–Hill.

Spicer, B. R. (1971). A pilot study of traffic conflicts at a rural dual carriageway intersection. RRL report, LR 410, Crowthorne.

Tamakoshi, T., Nakasu, K., Ishio, M., and Nakatani, S. (2004). Study on evaluation method of the traffic characteristics on highway bridges—bridge weigh-in-motion system. Technical note of National Institute of Land and Infrastructure Management, 188.

Underwood, R. T. (2006). *Road safety: Strategies and solutions.* Crows Nest, NSW: EA Books.

Uno, N., Iida, Y., Hiyoshi, K. and Arino, M. (2006). An Approach for Evaluating LOS from Viewpoint of Traffic Flow Safety Using Video Image Data, Proceedings of the 5th International Symposium on Highway Capacity and Quality of Service, pp. 189–198.

Zimmerman, G., Zimolong, B., and Erke, H. (1977). The development of the traffic conflicts technique in the Federal Republic of Germany. *Proceedings of 1st Workshop on Traffic Conflicts,* Oslo.

CHAPTER 7

Network Design for Freight Transport and Supply Chain

Tadashi Yamada

CONTENTS

7.1 Introduction	167
7.2 Discrete Network Design	168
7.2.1 Overview	168
7.2.2 Modeling Framework	170
7.3 Equilibrium Models for the Lower Level	171
7.3.1 TNE and SCNE	171
7.3.2 Example—Multiclass Multimodal TNE	173
7.3.3 Example—SC-T-SNE	174
7.4 Heuristic Approaches for Solving the Upper Level	178
7.4.1 Overview	178
7.4.2 Genetic Local Search	179
References	181

7.1 INTRODUCTION

There has been a growing appreciation of the importance of developing networks in the field of logistics and supply chain management (SCM), including freight transport network (FTN) and supply chain network (SCN). Freight transport constitutes an important activity undertaken between and within cities. An efficient spatial organization of multimodal transport systems has the potential to support the economic and sustainable development of cities, regions, and countries by alleviating various externalities such as traffic congestion, increased energy consumption, and negative environmental impacts.

Multimodal transport systems (and intermodal transport systems as well) are useful in expanding FTNs for cities, regions, and countries where much of the focus has been centered on road-based freight transport systems. In developing countries, the existing multimodal facilities consisting of roads, rails, and ports are still undeveloped, and road capacities, especially outside urban areas, are still inadequate. Furthermore, most of the port terminals provide very low levels of service due to the lack of berths and supporting equipment. There is therefore a need to develop transport infrastructures optimally to suit the needs for a well-coordinated and efficient multimodal operation.

"Supply chain" can be defined as a network of linkages between various economic entities for the passage of products from production to consumption, including manufacturers, wholesalers, retailers, freight carriers, and consumers. It also involves a variety of processes of manufacturing, order processing, transactions, transport, warehousing, inventory management, facility operation, and final pickup/delivery of goods. SCM has recently become a crucial long-term strategy for businesses, as consumer demands have become more diversified and international sales competition has intensified. The design of efficient SCN is crucial not only to businesses but also to transport agencies, because decisions on product distribution and freight transport are made practically by looking over the SCN as a whole. As such, identifying the mechanism of the development and expansion of SCN could facilitate the comprehension of how freight trips and flows generate in cities, regions, and countries.

There is a possibility of the SCN entities and traffic conditions on a transport network (TN) influencing each other's behaviors, since products are moved through the TN. Thus, the construction/renovation of TN links (i.e., TN design) has significant effect on the movement of the products and on the efficiency of the SCNs. The design of FTN and SCN should be undertaken with such an SCN-TN interaction being taken into account.

Recently, attention has also been paid on the development of reliable network when natural or man-made disaster happens (e.g., network vulnerability or network interdiction). However, for simplicity, this chapter only highlights the development of optimal FTN and SCN in terms of economic efficiency, even though the methodologies to be presented in this chapter can be applied to problems with objectives other than economic efficiency.

7.2 DISCRETE NETWORK DESIGN

7.2.1 Overview

If a problem involves the optimal actions of adding new links or selecting existing links for improvement in a network, it can be considered as a discrete network design problem (DNDP), which basically implies the selection of link additions to an existing transport network with given demand from each origin to each destination. The DNDP has been studied so far by Bruynooghe (1972), Steenbrink (1974), Poorzahedy and Turnquist (1982), Boyce (1984), Magnanti and Wong (1984), Chen and Alfa (1991), Yang and Bell (1998), Gao, Wu, and Sun (2005), Poorzahedy and Rouhani (2007), Ukkusuri and Patil (2009), Yamada et al. (2009), and Luathep et al. (2011).

Arnold, Peeters, and Thomas (2004) deal with the problem of optimally locating rail/road terminals for freight transport. Their approach is based on fixed-charge network design problems (e.g., Magnanti and Wong 1984; Balakrishnan, Magnanti, and Wong 1989), which are also applied by Melkote and Daskin (2001) for simultaneously optimizing facility locations and the design of the underlying transport network. Thus, network location theory, including the fixed-charge network design problems, location-routing problems (e.g., Laporte 1988; ReVelle and Laporte 1996; Min, Jayaraman, and Srivastava 1998), and hub location problems (e.g., O'Kelly 1987; Aykin 1990; Campbell 1994, 1996) has been a powerful tool for TN design and facility locations, even though it has not appropriately taken into account the change in traffic flow caused by expanding and improving TNs. Francis, McGinnis, and White (1992), Daskin (1995), and Drezner (1995) provide a comprehensive overview of network location models.

Luathep et al. (2011) indicates that DNDPs minimize (or maximize) a specific network performance index while accounting for route choice behavior in a user equilibrium (UE) or a stochastic user equilibrium (SUE) manner. This means that the change in traffic flow (and in product flow in the case of SCNs as well) caused by expanding and improving the networks is an important factor that should be taken into account in developing DNDP-based network design models for their practical use. The conventional network location models have not explicitly incorporated it on actual medium- or large-sized networks. In the design of FTN, traffic conditions on the transport network should be influenced by the network improvement actions implemented, especially when these actions are carried out on a larger scale as transport initiatives in interregional or urban transport planning. This allows the model to deal with more realistic situations. Therefore, such models have the potential to be used as a tool for strategic levels of multimodal freight transport planning, particularly in freight terminal development and FTN design. Yamada et al. (2009) deal with DNDP embedding the flow of freight or products, representing the route choice behavior in the TN in the context of UE.

7.2.2 Modeling Framework

The process of discrete network design can often be represented within the framework of bilevel programming (BP), because the decision maker at the upper level (e.g., transport agencies and SCN planners) may be able to influence the behavior of a decision maker at the lower level (e.g., SCN users of manufacturers, wholesalers, retailers, freight carriers, and consumers; and TN users of freight carriers and passengers). In cases where reader–follower relationships exist between the upper and lower level decision makers, BP-based network design models can be an effective tool for determining efficient logistics and freight transport initiatives. The BP-based model can generally be formulated as follows (e.g., Colson, Marcotte, and Savard 2005; Patriksson 2008):

$$\min_{x} f(x,v)$$

$$\text{s.t. } (x,v) \in Z \qquad (7.1)$$

$$v \in \arg\min_{z \in V(x)} \varphi(x,z)$$

where
f: upper level objective function ($f: R^n \times R^m \to R$)
x: upper level variables (i.e., decision variables or design variables) ($x \in R^n$)
v: lower level variables (i.e., state variables or responses) ($v \in R^m$)
Z: nonempty set of upper level and lower level variables ($Z \subseteq R^m \times R^m$)
$V(x)$: closed and convex set of the lower level optimization problem ($V(x) \subseteq R^m$)
$\phi(x,\bullet)$: differentiable function ($\phi(x,\bullet): R^m \to R$)

Bard (1998) and Dempe (2002) described the BP in detail. In the case of equilibrium models incorporated within the lower level, this type of problem also involves a mathematical program with equilibrium constraints (MPEC) (e.g., Luo, Pang, and Ralph 1996; Outrata, Kocvara, and Zowe 1998). The MPEC can generally be formulated as follows:

$$\min_{x} f(x,v)$$

$$\text{s.t. } (x,v) \in Z \qquad (7.2)$$

$$v \in S(x)$$

where $S(x)$ is the set of $v(\in V(x))$.

$S(x)$ denotes the solution to the following variational inequality (VI) problem (i.e., lower level problem), parameterized by the upper level variable x:

$$\langle F(x,v), z-v \rangle \geq 0 \quad \forall z \in V(x) \tag{7.3}$$

where $F(x,\bullet)$ is vector valued mapping $(F(x,\cdot): R^m \to R^m)$.

The VI problem also indicates that the inclusion $v \in S(x)$ holds if and only if

$$-F(x,v) \in N_{V(x)}(v) \tag{7.4}$$

where $N_{V(x)}(v)$ is the normal cone to set $V(x)$ at $v \in V(x)$.

If $F(x,\bullet) = \nabla \varphi(x,\bullet)$—that is, if the mapping F is given as a gradient mapping $\nabla \varphi(x,\bullet)$—Equation (7.4) denotes the first-order optimality conditions for the parameterized optimization problem to

$$\min_{v \in V(x)} \varphi(x,v) \tag{7.5}$$

The BP can likewise be considered as a special case of MPEC obtained when the lower level VI reduces to the optimality conditions for an optimization problem. The mapping $x \mapsto S(x)$ can represent Karush-Kuhn-Tucker (KKT) conditions for the optimization problem (Equation 7.2) under a constraint qualification.

In addition, when $V(x)$ ($v \in V(x)$) is the non-negative orthant R_+^n for all x in VI (Equation 7.3), the parametric VI reduces to a parametric complementarity problem. Thus, the optimization problem (Equation 7.2) is equivalent to the following mathematical problem with complementarity constraints (MPCC) (e.g., Lin and Fukushima 2003):

$$\min_x f(x,v)$$

$$\text{s.t. } (x,v) \in Z \tag{7.6}$$

$$v \geq 0, \quad F(x,v) \geq 0, \quad \langle F(x,v), v \rangle = 0$$

7.3 EQUILIBRIUM MODELS FOR THE LOWER LEVEL

7.3.1 TNE and SCNE

Equilibrium models have typically been used to represent the behavior of network users being incorporated within the lower level. There have been a variety of equilibrium models developed and applied in transport (e.g., Bell and Iida 1995; Nagurney 2000), economics (e.g., Arrow and Intrilligator 1981; Holguín-Veras 2000), and finance (e.g., Nagurney 1999). As such, the equilibrium models have a long tradition and typically relate to network economics.

TN equilibrium (TNE) models are typical procedures being used to represent travelers' behavior in a transport network. Patriksson (1994), Florian and Hearn (1995), and Boyce (2007) provide a comprehensive overview of such models. TNE originated with the Wardrop equilibrium (i.e., UE) by Wardrop (1952) and basically deals with fixed origin–destination (OD) traffic demands, which were first formulated by Jorgensen (1963) and called "user equilibrium traffic assignment with fixed origin–destination flows." In the case of OD traffic demands varying in response to the service levels of the TN, the method is called "user equilibrium traffic assignment with elastic demands," which Beckmann, McGuire, and Winsten (1955) typically formulated. If the method considers multiple types of users simultaneously, it is termed "multiclass UE" (e.g., Yang and Bell 1998).

The steady behaviors of decision makers in SCNs are also characterized by a group of equilibrium conditions. The equilibrium models can therefore be the tools for investigating their behaviors. SCN equilibrium (SCNE) models describe what happens on multitiered SCNs, incorporating decentralized decisions made by multiple entities on the SCNs and their behavioral interaction. Nagurney, Dong, and Zhang (2002) first developed an SCNE model incorporating interaction and correlation in the business process of a product flowing from manufacturers to consumers via retailers. The equilibrium conditions are described with VIs, allowing the behavior of three decision makers in a decentralized SCN to be taken into account. The SCNE model was then expanded to consider random consumers' demand (Dong, Zhang, and Nagurney 2004), electronic commerce (Nagurney et al. 2005; Zhao and Nagurney 2008), reverse supply chains (Nagurney and Toyasaki 2005; Hammond and Beullens 2007; Yang, Wang, and Li 2009), corporate social responsibility (Cruz 2008; Cruz and Matsypura 2009), production capacity constraints (Meng, Huang, and Cheu 2009; Hamdouch 2011), behavior of raw material suppliers (Yang et al. 2009; Cruz and Liu 2011), competition among SCNs (Nagurney 2010; Rezapour and Farahani 2010; Nagurney and Yu 2012), multiperiod (Cruz and Liu 2011; Hamdouch 2011), and dynamic aspects (Daniele 2010). Furthermore, the SCNE models have also been proved to be able to be reformulated and solved as TNE problems (Nagurney 2006).

Harker and Friesz (1985, 1986a, 1986b) formulate a general predictive network equilibrium model of freight transport systems, which is called the generalized spatial equilibrium model. This model takes into account the behaviors of freight carriers and shippers. Fernández, de Cea, and Alexandra (2003) also adopt the simultaneous demand–supply network equilibrium approach to represent the behavior of freight carriers and shippers. This model only uses a single network and integrates the shipper–carrier interaction in a simultaneous mathematical formulation. Aggregate-type multimodal freight transport network models have also been proposed on the basis of strategic freight network planning (e.g., Friesz, Tobin, and Harker 1983; Crainic, Florian, and Leal 1990; Guelat, Florian, and Crainic 1990; Tavasszy 1996; Southworth and Peterson 2000; Yamada et al. 2009).

Having further developed the SCNE model by integrating SCNs with a TN, Yamada et al. (2011) propose a supply chain–transport supernetwork equilibrium (SC-T-SNE) model. In this model, the behaviors of six entities within a supply chain–transport supernetwork, including manufacturers, wholesalers, retailers, freight carriers, demand markets, and TN users, are described. With the behavior of TN users including freight vehicles being incorporated, the model allows for endogenously determining transport costs based on freight carriers' decision making, as well as for investigating mutual effects between behavioral changes in SCNs and the TN. Notably, the effects of traffic conditions in the road network on the behavior of the entities on each SCN and vice versa are explored. To enhance the applicability of the model, the behavior of wholesalers and the facility costs for manufacturers, wholesalers, retailers, and freight carriers are also embedded within the model; these are not taken into consideration in the existing SCNE models. Furthermore, the SC-T-SNE model has the capacity of elucidating the relationship between traffic flow and product movement throughout the entire SCN. In that sense, the SC-T-SNE model is more applicable to SCNs than currently existing TN models (e.g., Friesz et al. 1983; Harker and Friesz 1986a, 1986b; Crainic et al. 1990; Guelat et al. 1990; Fernandez et al. 2003; Yamada et al. 2009).

7.3.2 Example—Multiclass Multimodal TNE

Yamada et al. (2009) utilize a multiclass multimodal TNE, a modal split-assignment model, to represent freight and passenger flows in a TN. The model allows freight and passengers to be treated as multiclass users, with modal split and route choice carried out simultaneously by converting the multimodal network into a unimodal abstract-mode network. A detailed representation of freight movement within terminals is developed based on network descriptions proposed by Guelat et al. (1990), Tavasszy (1996), and Southworth and Peterson (2000), incorporating loading/unloading, storage, and administrative processes. This can be formulated as follows:

Find $w_a^{b*} \in \kappa$ such that

$$\sum_{b=1}^{P}\sum_{a\in A} c_a^b(\tilde{w}^*) \times (w_a^b - w_a^{b*}) \geq 0 \quad \forall \tilde{w} \in \kappa \quad (7.7)$$

where

w_a^{b*}: the user equilibrium flow of link a for user type b
$\tilde{w}$: the P-dimensional column vector with the components $\{w_a^1, ..., w_a^P\}$
A: the set of links on the transport network
$c_a^b(\bullet)$: the generalized cost on link a for user type i
κ: the nonempty set defined as $\kappa \equiv \{\tilde{w}|$ satisfying the non-negative path flows and conservation of flow$\}$

Inequality (7.7) can be solved by the diagonalization method (e.g., Florian and Spiess 1982; Sheffi 1985; Thomas 1991). The generalized link costs are described to be made up of a fare component and a time cost component. The time spent on the link includes travel time, waiting time, and loading/unloading time, depending on the type of links. Link types are classified into link ways and terminal links for each mode. Although the fare component is assumed to have a fixed value and does not depend on volume, the time cost component, particularly the time spent on the link, is a function of volume and differs by link type. The time spent on the link is represented by a continuous function in the form of polynomial approximation presented by Crainic et al. (1990) to keep the link cost function monotonically increasing.

7.3.3 Example—SC-T-SNE

In case the MPEC-based DNDP is developed in terms of SCN efficiency, the SC-T-SNE (Yamada et al. 2011) can be used within the lower level. The behavior of six entities in a supply chain transport supernetwork is incorporated within the SC-T-SNE model, including manufacturers, wholesalers, retailers, freight carriers, consumers (i.e., demand markets), and transport network users (see Yamada et al. 2011, for the details of the mathematical formulation). Here, equilibrium solutions are denoted by "*" and the equilibrium conditions governing the SC-T-SNE

model are equivalent to the solution to the VI given by: determine
$(Q^{1*}, Q^{2*}, Q^{3*}, \gamma^*, \delta^*, \rho^{4*}, X^*, c_{rs}^*) \in R_+^{H^1+H^2+H^3+nY+oY+LY+e^5e^6e^4+e^5e^6}$, satisfying:

$$\sum_{y=1}^{Y}\sum_{h^y=1}^{u^y}\sum_{i^y=1}^{m^y}\sum_{j^y=1}^{n^y}\sum_{p^{1y}\in E^{1y}} \left[\frac{\partial B_{i^y}(Q^{1y*})}{\partial q_{h^y i^y j^y}^{p^{1y}}} + \frac{\partial g_{i^y}(Q^{1y*})}{\partial q_{h^y i^y j^y}^{p^{1y}}} + \frac{\partial c_{i^y j^y}(Q^{1y*})}{\partial q_{h^y i^y j^y}^{p^{1y}}} \right.$$

$$+ \frac{\partial c_{j^y}(Q^{1y*})}{\partial q_{h^y i^y j^y}^{p^{1y}}} + \frac{\partial g_{j^y}(Q^{1y*})}{\partial q_{h^y i^y j^y}^{p^{1y}}} + \frac{\partial g_{h^y}(Q^{1y*}, Q^{2y*}, Q^{3y*})}{\partial q_{h^y i^y j^y}^{p^{1y}}}$$

$$+ C_{h^y i^y j^y}^{p^{1y}}(Q^{1*}, Q^{2*}, Q^{3*}, X^*) + q_{h^y i^y j^y}^{p^{1y}} \frac{\partial C_{h^y i^y j^y}^{p^{1y}}(Q^{1*}, Q^{2*}, Q^{3*}, X^*)}{\partial q_{h^y i^y j^y}^{p^{1y}}}$$

$$+ \sum_{j^y=1}^{n^y}\sum_{k^y=1}^{o^y}\sum_{p^{2y}\in E^{2y}} q_{h^y j^y k^y}^{p^{2y}} \frac{\partial C_{h^y j^y k^y}^{p^{2y}}(Q^{1*}, Q^{2*}, Q^{3*}, X^*)}{\partial q_{h^y i^y j^y}^{p^{1y}}}$$

$$\left. + \sum_{k^y=1}^{o^y}\sum_{l=1}^{L}\sum_{p^{3y}\in E^{3y}} q_{h^y k^y l}^{p^{3y}} \frac{\partial C_{h^y k^y l}^{p^{3y}}(Q^{1*}, Q^{2*}, Q^{3*}, X^*)}{\partial q_{h^y i^y j^y}^{p^{1y}}} - \gamma_{j^y}^* \right]$$

$$\times \left[q_{h^y i^y j^y}^{p^{1y}} - q_{h^y i^y j^y}^{p^{1y*}} \right]$$

$$+ \sum_{y=1}^{Y}\sum_{h^y=1}^{u^y}\sum_{j^y=1}^{n^y}\sum_{k^y=1}^{o^y}\sum_{p^{2y}\in E^{2y}} \left[\frac{\partial c_{k^y}(Q^{2y*})}{\partial q_{h^y j^y k^y}^{p^{2y}}} + \frac{\partial g_{k^y}(Q^{2y*})}{\partial q_{h^y j^y k^y}^{p^{2y}}} + \frac{\partial c_{j^y k^y}(Q^{2y*})}{\partial q_{h^y j^y k^y}^{p^{2y}}} \right.$$

$$+ \frac{\partial g_{h^y}(Q^{1y*}, Q^{2y*}, Q^{3y*})}{\partial q_{h^y j^y k^y}^{p^{2y}}} + C_{h^y j^y k^y}^{p^{2y}}(Q^{1*}, Q^{2*}, Q^{3*}, X^*)$$

$$+ q_{h^y j^y k^y}^{p^{2y}} \frac{\partial C_{h^y j^y k^y}^{p^{2y}}(Q^{1*}, Q^{2*}, Q^{3*}, X^*)}{\partial q_{h^y j^y k^y}^{p^{2y}}}$$

$$+ \sum_{i^y=1}^{m^y}\sum_{j^y=1}^{n^y}\sum_{p^{1y}\in E^{1y}} q_{h^y i^y j^y}^{p^{1y}} \frac{\partial C_{h^y i^y j^y}^{p^{1y}}(Q^{1*}, Q^{2*}, Q^{3*}, X^*)}{\partial q_{h^y j^y k^y}^{p^{2y}}}$$

$$+ \sum_{k^y=1}^{o^y}\sum_{l=1}^{L}\sum_{p^{3y}\in E^{3y}} q_{h^y k^y l}^{p^{3y}} \frac{\partial C_{h^y k^y l}^{p^{3y}}(Q^{1*}, Q^{2*}, Q^{3*}, X^*)}{\partial q_{h^y j^y k^y}^{p^{2y}}}$$

$$\left. + \gamma_{j^y}^* - \delta_{k^y}^* \right] \times \left[q_{h^y j^y k^y}^{p^{2y}} - q_{h^y j^y k^y}^{p^{2y*}} \right]$$

$$+ \sum_{y=1}^{Y}\sum_{h^y=1}^{u^y}\sum_{k^y=1}^{o^y}\sum_{l=1}^{L}\sum_{p^{3y}\in E^{3y}}\left[\frac{\partial c_{k^y l}(Q^{3y*})}{\partial q_{h^y k^y l}^{p^{3y}}} + \frac{\partial g_{h^y}(Q^{1y*},Q^{2y*},Q^{3y*})}{\partial q_{h^y k^y l}^{p^{3y}}}\right.$$

$$+ C_{h^y k^y l}^{p^{3y}}(Q^{1*},Q^{2*},Q^{3*},X^*) + q_{h^y k^y l}^{p^{3y}}\frac{\partial C_{h^y k^y l}^{p^{3y}}(Q^{1*},Q^{2*},Q^{3*},X^*)}{\partial q_{h^y k^y l}^{p^{3y}}}$$

$$+ \sum_{i^y=1}^{m^y}\sum_{j^y=1}^{n^y}\sum_{p^{1y}\in E^{1y}} q_{h^y i^y j^y}^{p^{1y}}\frac{\partial C_{h^y i^y j^y}^{p^{1y}}(Q^{1*},Q^{2*},Q^{3*},X^*)}{\partial q_{h^y k^y l}^{p^{3y}}}$$

$$+ \sum_{j^y=1}^{n^y}\sum_{k^y=1}^{o^y}\sum_{p^{2y}\in E^{2y}} q_{h^y j^y k^y}^{p^{2y}}\frac{\partial C_{h^y j^y k^y}^{p^{2y}}(Q^{1*},Q^{2*},Q^{3*},X^*)}{\partial q_{h^y k^y l}^{p^{3y}}}$$

$$+ \delta_{k^y}^* - \rho_l^{4y*}\left.\right] \times \left[q_{h^y k^y l}^{p^{3y}} - q_{h^y k^y l}^{p^{3y*}}\right]$$

$$+ \sum_{y=1}^{Y}\sum_{j^y=1}^{n^y}\left[\sum_{h^y=1}^{u^y}\left(\sum_{i^y=1}^{m^y}\sum_{p^{1y}\in E^{1y}} q_{h^y i^y j^y}^{p^{1y*}} - \sum_{k^y=1}^{o^y}\sum_{p^{2y}\in E^{2y}} q_{h^y j^y k^y}^{p^{2y*}}\right)\right] \times \left[\gamma_{j^y} - \gamma_{j^y}^*\right]$$

$$+ \sum_{y=1}^{Y}\sum_{k^y=1}^{o^y}\left[\sum_{h^y=1}^{u^y}\left(\sum_{j^y=1}^{n^y}\sum_{p^{2y}\in E^{2y}} q_{h^y j^y k^y}^{p^{2y*}} - \sum_{l=1}^{L}\sum_{p^{3y}\in E^{3y}} q_{h^y k^y l}^{p^{3y*}}\right)\right] \times \left[\delta_{k^y} - \delta_{k^y}^*\right]$$

$$+ \sum_{y=1}^{Y}\sum_{l=1}^{L}\left[\sum_{h^y=1}^{u^y}\sum_{k^y=1}^{o^y}\sum_{p^{3y}\in E^{3y}} q_{h^y k^y l}^{p^{3y*}} - d_l^y(\rho^{4y*})\right] \times \left[\rho_l^{4y} - \rho_l^{4y*}\right]$$

$$+ \sum_{r\in G}\sum_{s\in A}\sum_{p_{rs}\in E_{rs}}\left[t_{rs}^{p_{rs}}(Q^{1*},Q^{2*},Q^{3*},X^*) - c_{rs}^*\right] \times \left[w_{rs}^{p_{rs}} - w_{rs}^{p_{rs}*}\right]$$

$$+ \sum_{r\in G}\sum_{s\in A}\left[\sum_{p_{rs}\in E_{rs}} w_{rs}^{p_{rs}*} - d_{rs}(c_{rs}^*)\right] \times \left[c_{rs} - c_{rs}^*\right] \geq 0$$

$$\forall \left(Q^1, Q^2, Q^3, \gamma, \delta, \rho^4, X, c_{rs}\right) \in R_+^{H^1+H^2+H^3+nY+oY+LY+e^5e^6+e^5e^6} \quad (7.8)$$

where

$q_{h^y i^y j^y}^{p^{1y}}$: the amount of product y transacted/transported from manufacturer i^y to wholesaler j^y by freight carrier h^y using path p^{1y}

$q_{h^y j^y k^y}^{p^{2y}}$: the amount of product y transacted/transported between wholesaler j^y to retailer k^y by freight carrier h^y using path p^{2y}

$q_{h^y k^y l}^{p^{3y}}$: the amount of product y transacted/transported from retailer k^y to demand market l by freight carrier h^y using path p^{3y}

Q^{1y}: the $u^y m^y n^y e^{1y}$-dimensional vector with component $h^y i^y j^y p^{1y}$ denoted by $q_{h^y i^y j^y}^{p^{1y}}$ representing shipments of product y between manufacturers and wholesalers

Q^{2y}: the $u^y n^y o^y e^{2y}$-dimensional vector with component $h^y j^y k^y p^{2y}$ denoted by $q_{h^y j^y k^y}^{p^{2y}}$ representing shipments of product y between wholesalers and retailers

Q^{3y}: the $u^y o^y L e^y$-dimensional vector with component $h^y k^y l p^{3y}$ denoted by $q_{h^y k^y l}^{p^{3y}}$ representing shipments of product y between retailers and demand markets

Q^1: the H^1-dimensional vector with components: $Q^{11},\ldots, Q^{1Y}$, where $H^1 \left(= \sum_{y=1}^{Y} \left(u^y m^y n^y e^{1y}\right)\right)$

Q^2: the H^2-dimensional vector with components: $Q^{21},\ldots, Q^{2Y}$, where $H^2 \left(= \sum_{y=1}^{Y} \left(u^y n^y o^y e^{2y}\right)\right)$

Q^3: the H^3-dimensional vector with components: $Q^{31},\ldots, Q^{3Y}$, where $H^3 \left(= \sum_{y=1}^{Y} \left(u^y o^y L e^{3y}\right)\right)$

$B_{i^y}(\bullet)$: the production cost to manufacturer i^y for product y

$g_{i^y}(\bullet)$: the facility cost to manufacturer i^y

$g_{j^y}(\bullet)$: the facility cost to wholesaler j^y

$g_{k^y}(\bullet)$: the facility cost to retailer k^y

$g_{h^y}(\bullet)$: the facility cost to freight carrier h^y

$c_{j^y}(\bullet)$: the handling/inventory costs to wholesaler j^y

$c_{k^y}(\bullet)$: the handling/inventory costs to retailer k^y

$c_{i^y j^y}(\bullet)$: the transaction cost for product y incurred between manufacturer i^y and wholesaler j^y (excluding transport cost incurred between i^y and j^y)

$c_{j^y k^y}(\bullet)$: the transaction cost for product y incurred between wholesaler j^y and retailer k^y (excluding transport cost incurred between j^y and k^y)

$c_{k^y l}(\bullet)$: the transaction cost for product y incurred between retailer k^y and demand market l (excluding transport cost incurred between k^y and l)

C_{rs}: the travel cost incurred between $r(\in G)$ and $s(\in A)$

$C_{h^y i^y j^y}^{p^{1y}}(\bullet)$: the unit operation cost (per transport volume) of freight carrier h^y for transporting products y from manufacturer i^y to wholesaler j^y using path p^{1y}

$C_{h^y j^y k^y}^{p^{2y}}(\bullet)$: the unit operation cost (per transport volume) of freight carrier h^y for transporting products y from wholesaler j^y to retailer k^y using path p^{2y}

$C_{h^y k^y l}^{p^{3y}}(\bullet)$: the unit operation cost (per transport volume) of freight carrier h^y for transporting products y from retailer k^y to demand market l using path p^{3y}

X: the $e^5 e^6 e^4$-dimensional vector with component rsp_{rs} denoted by $w_{rs}^{p_{rs}}$

$w_{rs}^{p_{rs}}$: the traffic volume of passenger cars traveling between r and s using path p_{rs}

ρ_l^{4y}: the market price of product y at demand market l

ρ^{4y}: L-dimensional vector for product y with component l denoted by ρ_l^{4y}

ρ^4: LY-dimensional vector for product y with component ly denoted by ρ_l^{4y}

$d_l^y(\bullet)$: the demand function of product y at demand market l

e^4: the number of path nodes between r and s

e^5: the number of origin nodes for passenger cars

e^6: the number of destination nodes for passenger cars

$t_{rs}^{p_{rs}}(\bullet)$: the travel time on path p_{rs}

$d_{rs}(\bullet)$: the traffic demand function between r and s

$\gamma_{jy}, \delta_{k^y}$: the Lagrange multiplier

γ: the nY-dimensional vector with component jy denoted by γ_{jy}

δ: the oY-dimensional vector with component ky denoted by δ_{k^y}

The existence and uniqueness of the solution of VI (7.8) are proved by Yamada et al. (2011). The VI can be solved using the procedures shown in Yamada et al. (2011), which were originally proposed by Meng et al. (2007).

7.4 HEURISTIC APPROACHES FOR SOLVING THE UPPER LEVEL

7.4.1 Overview

The advantage of applying heuristic techniques to BP or MPEC is that such techniques can handle complex problems and provide the flexibility of the design of such problems if applied as optimization techniques. The BP-based or MPEC-based DNDP typically involves a combinatorial optimization problem. These techniques can also compute approximately efficient solutions in relatively shorter times. Hence, in the past decade, several metaheuristics have been developed and applied in the field of soft computing. A vital role of these techniques is to solve complex and difficult mathematical programming problems, which often involve NP-hard problems. These cannot ensure obtaining exact optimal solutions, but can provide reasonable and practical solutions. Consequently, these have been commonly applied to combinatorial optimization problems wherein exact optimal solutions are hard to determine.

Ribeiro and Hansen (2001), Michalewicz and Fogel (2002), Glover and Kochenberger (2003), Herz and Widmer (2003), and Resende and Pinho de Sousa (2004) provided excellent introductions and basic concepts of metaheuristics. In general, genetic algorithms (GAs) (e.g., Goldberg 1989), tabu search (TS) (e.g., Glover and Laguna 1997), simulated annealing (SA) (e.g., Aarts and Korst 1989), and ant colony optimization (ACO) (e.g., Dorigo and Stutzle 2004) are typical solution techniques in metaheuristics.

7.4.2 Genetic Local Search

Yamada et al. (2009) applied genetic local search (GLS) to solve the upper level of the MPEC-based DNDP for determining a suitable set of TN improvement actions from a number of possible actions, such as the renovation of existing infrastructure and the establishment of new roads, railways, sea links, and freight terminals. GLS, sometimes called memetic algorithms or hybrid genetic algorithms, is a hybrid metaheuristic technique combining genetic algorithms with local search (e.g., Ackley 1987; Radcliffe and Surry 1994; Jaszkiewicz 2002; Jaszkiewicz and Kominek 2003; Arroyo and Armentano 2005). A possible explanation of the superiority of GLS is that it offers the potential to cover the weakness of GAs in searching local areas, because GAs can efficiently find the vicinity of the optimal solution from a wider range. Thus, the heuristics on the basis of the GLS scheme often outperform other metaheuristics on combinatorial optimization problems (e.g., Murata and Ishibuchi 1994; Merz and Freisleben 1997; Galinier and Hao 1999). The GLS algorithm can be described as follows (Yamada et al. 2009):

Parameters:
    Size of the current population: M
    Stopping criterion: predefined number of generations (i.e., number of iterations given in advance)
Initialization:
    Current population $x := \emptyset$
    Current generation $\alpha = 0$
    Repeat M times
        Generate a feasible solution (i.e., an individual) x_a by a randomized algorithm
        Add x_a to x for constructing an initial population

Main loop:
> **Repeat**
>> Select K best individuals to make x_1 (i.e., elitist selection)
>> Evaluate x and select $(M-K)$ individuals according to their fitness to obtain x_2
>> Draw two solutions $(M-K)$ times from x_2 as parent solutions
>> Generate $(M-K)$ child solutions (i.e., offspring) to obtain x_3 by applying uniform crossover procedures with pre-defined crossover rates
>> Mutate each bit of $(M-K)$ child solutions in x_3 with some low probability to obtain y_4
>> Apply local search to x_4 to obtain x_5
>> Develop current population x consisting of x_1 and x_5
>> Change α to $\alpha + 1$
>
> **Until** the stopping criterion is met

When the size of TN is relatively large and the search space is relatively small, Yamada et al. (2009) revealed that GLS can offer the best performance among GA-based and TS-based procedures, considering its robustness and faster searching ability. The GA-based procedures include the basic version of the GA-based scheme (i.e., simple GA) with standard operators in selection and reproduction, crossover and mutation, and the improved version with additive operators, such as elitist selection (e.g., Goldberg 1989) used to preserve some of the best individuals for further generation and uniform crossover (e.g., Syswerda 1989). TS-based procedures are TS-B, TS-MSM, and TS-MSM&LM. TS-B is the fundamental version of TS starting with a feasible solution being generated randomly, where the modifications of the current solution are repetitively examined by the process of move, similar to the local search incorporated within GLS. To avoid cycling, the simplest form of adaptive memory, that is, a tabu list (tabu restrictions or short-term memory), is generally utilized, with its tabu tenure (i.e., size of the tabu list) being defined in advance. A simple aspiration criterion is also included to remove the tabu status of a solution if the value of the solution is better than that of the current best solution. TS-MSM incorporates an intensification process (Rochat and Taillard 1995) by restarting from high-quality solutions within TS-B. TS-MSM&LM embeds the long-term memory (e.g., Diaz and Fernandez 2001) within TS-MSM for exploiting features historically found desirable.

Yamada et al. (2009) show an example of the MPEC-based DNDP for investment planning in developing multimodal freight transport networks, which can be translated to a problem of identifying and selecting a suitable set of TN improvement actions from a number of possible actions. The lower level describes the multimodal multiuser equilibrium flow on the TN, whereas the upper level determines the best combination of the actions such that the benefit-cost ratio to identify the economic effectiveness of freight network improvement actions is maximized. The model is applied to a large-scale transport network in the Philippines to investigate a possible development strategy for improving the interregional freight transport network.

REFERENCES

Aarts, E. H. L., and Korst, J. H. M. (1989). *Simulated annealing and Boltzmann machines: A stochastic approach to combinatorial optimization and neural computing.* Chichester, UK: John Wiley & Sons.

Ackley, D. H. (1987). *A connectionist machine for genetic hill climbing.* Boston, MA: Kluwer Academic Publishers.

Arnold, P., Peeters, D., and Thomas, I. (2004). Modeling a rail/road intermodal transportation system. *Transportation Research Part E* 40: 255–270.

Arrow, K. J., and Intrilligator, M. D., eds. (1981). *Handbook of mathematical economics.* Amsterdam, The Netherlands: North-Holland Publishing.

Arroyo, J. E. C., and Armentano, V. A. (2005). Genetic local search for multiobjective flow shop scheduling problems. *European Journal of Operational Research* 151: 717–738.

Aykin, T. (1990). On a quadratic integer program for the location of interacting hub facilities. *European Journal of Operational Research* 46: 409–411.

Balakrishnan, A., Magnanti, T. L., and Wong, R. T. (1989). A dual-ascent procedure for large-scale uncapacitated network design. *Operations Research* 37: 716–740.

Bard, J. F. (1998). *Practical bilevel optimization: Algorithms and applications.* Dordrecht, The Netherlands: Kluwer Academic Publishers.

Beckmann, M., McGuire, C. B., and Winsten, C. B. (1955). *Studies in the economics of transportation.* New Haven/Santa Monica, CA: Yale University Press/Rand Corporation (published as RM-1488).

Bell, M. G. H., and Iida, Y. (1997). *Transportation network analysis.* Chichester, UK: Wiley.

Boyce, D. E. (1984). Urban transportation network equilibrium and design models: Recent achievements and future prospective. *Environment and Planning* 16A: 1445–1474.

——— (2007). Forecasting travel on congested urban transportation networks: review and prospects for network equilibrium models. *Networks and Spatial Economics* 7: 99–128.

Bruynooghe, M. (1972). An optimal method of choice of investments in a transport network. Presentation at Planning & Transport Research & Computation Seminars on Urban Traffic Model Research, London, UK.

Campbell, J. (1994). Integer programming formulations of discrete hub location problems. *European Journal of Operational Research* 72: 387–405.

——— (1996). Hub location and the p-hub median problem. *Operations Research* 44: 923–935.

Chen, M., and Alfa, A. S. (1991). A network design algorithm using a stochastic incremental traffic assignment approach. *Transportation Science* 25: 215–224.

Colson, B., Marcotte, P., and Savard, G. (2005). Bilevel programming: A survey. *Quarterly Journal of Operations Research* 3: 87–107.

Crainic, T. G., Florian, M., and Leal, J. (1990). A model for the strategic planning of national freight transportation by rail. *Transportation Science* 24: 1–24.

Cruz, J. M. (2008). Dynamics of supply chain networks with corporate social responsibility through integrated environmental decision making. *European Journal of Operational Research* 184: 1005–1031.

Cruz, J. M., and Liu, Z. (2011). Modeling and analysis of the multiperiod effects of social relationship on supply chain networks. *European Journal of Operational Research* 214: 39–52.

Cruz, J. M., and Matsypura, D. (2009). Supply chain networks with corporate social responsibility through integrated environmental decision making. *International Journal of Production Research* 47: 621–648.

Daniele, P. (2010). Evolutionary variational inequalities and applications to complex dynamic multilevel models. *Transportation Research Part E* 46: 855–880.

Daskin, M. S. (1995). *Network and discrete location: Models, algorithms, and applications*. New York: John Wiley & Sons.

Dempe, S. (2002). *Foundations of bilevel programming*. Dordrecht, The Netherlands: Kluwer Academic Publishers.

Diaz, J. A., and Fernandez, E. (2001). A tabu search heuristic for the generalized assignment problem. *European Journal of Operational Research* 132: 22–38.

Dong, J., Zhang, D., and Nagurney, A. (2004). A supply chain network equilibrium model with random demands. *European Journal of Operational Research* 156: 194–212.

Dorigo, M., and Stutzle, T. (2004). *Ant colony optimization*. Boston, MA: MIT Press.

Drezner, Z. (1995). *Facility location: A survey of applications and methods*. Heidelberg, Germany: Springer–Verlag.

Fernández, J. E. L., de Cea, J. Ch., and Alexandra, S. O. (2003). A multimodal supply–demand equilibrium model for predicting intercity freight flows. *Transportation Research Part B* 37: 615–640.

Florian, M., and Hearn, D. (1995). Network equilibrium models and algorithms. In *Network routing: Handbooks in operations research and management science,* ed. Ball, M. O., Magnanti, T. L., Monma, C. L., and Nemhauser, G. L., 485–550. Amsterdam, The Netherlands: Elsevier Science.

Florian, M., and Spiess, H. (1982). The convergence of diagonalization algorithms for asymmetric network equilibrium problems. *Transportation Research Part B* 16: 477–483.

Francis, R. L., McGinnis, L. F., and White, J. A. (1992). *Facility layout and location: An analytical approach.* Upper Saddle River, NJ: Prentice Hall.

Friesz, T. L., Tobin, R. L., and Harker, P. T. (1983). Predictive intercity freight network models: The state of the art. *Transportation Research Part A* 17: 409–417.

Galinier, P., and Hao, J. K. (1999). Hybrid evolutionary algorithms for graph coloring. *Combinatorial Optimization* 3: 379–397.

Gao, Z., Wu, J., and Sun, H. (2005). Solution algorithm for the bi-level discrete network design problem. *Transportation Research Part B* 39: 479–495.

Glover, F., and Kochenberger, G. A. (2003). *Handbook of metaheuristics.* Boston, MA: Kluwer Academic Publishers.

Glover, F., and Laguna, M. (1997). *Tabu search.* Boston, MA: Kluwer Academic Publishers.

Goldberg, D. E. (1989). *Genetic algorithms in search, optimization, and machine learning.* Reading, MA: Addison Wesley.

Guelat, J., Florian, M., and Crainic, T. G. (1990). A multimode multiproduct network assignment model for strategic planning of freight flows. *Transportation Science* 24: 25–39.

Hamdouch, Y. (2011). Multiperiod supply chain network equilibrium with capacity constraints and purchasing strategies. *Transportation Research Part C* 19: 803–820.

Hammond, D., and Beullens, P. (2007). Closed-loop supply chain network equilibrium under legislation. *European Journal of Operation Research* 183: 895–908.

Harker, P. T., and Friesz, T. L. (1985). The use of equilibrium network models in logistics management: With application to the US coal industry. *Transportation Research Part B* 19B: 457–470.

——— (1986a). Prediction of intercity freight flow, I: Theory. *Transportation Research Part B* 20B: 139–153.

——— (1986b). Prediction of intercity freight flow, II: Mathematical formulations. *Transportation Research Part B* 20B: 155–174.

Herz, A., and Widmer, M. (2003). Guidelines for the use of metaheuristics in combinatorial optimization. *European Journal of Operational Research* 151: 247–252.

Holguín-Veras, J. (2000). A framework for an integrative freight market simulation. *Proceedings of IEEE 3rd Annual Intelligent Transportation System Conference*, 476–481.

Jaszkiewicz, A. (2002). Genetic local search for multiobjective combinatorial optimization. *European Journal of Operational Research* 137:50–71.

Jaszkiewicz, A., and Kominek, P. (2003). Genetic local search with distance preserving recombination operator for a vehicle routing problem. *European Journal of Operational Research* 151:352–364.

Jorgensen, N. O. (1963). Some aspects of the urban traffic assignment problem. Master's thesis, Institute of Transportation and Traffic Engineering, University of California, Berkeley.

Laporte, G. (1988). Location-routing problems. In *Vehicle routing: Methods and studies,* ed. Golden, B. L. and A. A. Assad, 163–198. Amsterdam, The Netherlands: North-Holland.

Lin, G. H., and Fukushima, M. (2003). New relaxation method for mathematical programs with complementarity constraints. *Journal of Optimization Theory and Application* 118:81–116.

Luathep, P., Sumalee, A., Lam, W. H. K., Li, Z.-C., and Lo, H. K. (2011). Global optimization method for mixed transportation network design problem: A mixed-integer linear programming approach. *Transportation Research Part B* 45:808–827.

Luo, J. S., Pang, Z. Q., and Ralph, D. (1996). *Mathematical programs with equilibrium constraints.* Cambridge, UK: Cambridge University Press.

Magnanti, T. L., and Wong, R. T. (1984). Network design and transportation planning: Models and algorithms. *Transportation Science* 18:1–55.

Melkote, S., and Daskin, M. S. (2001). An integrated model of facility location and transportation network design. *Transportation Research Part A* 35:515–538.

Meng, Q., Huang, Y., and Cheu, R. L. (2007). A note on supply chain network equilibrium models. *Transportation Research Part E* 43:60–71.

——— (2009). Competitive facility location on decentralized supply chains. *European Journal of Operational Research* 196:487–499.

Merz, P., and Freisleben, B. (1997). Genetic local search for the TSP: New results. In *Proceedings 1997 IEEE International Conference on Evolutionary Computation.* New York: IEEE Press, 159–164.

Michalewicz, Z., and Fogel, D. B. (2002). *How to solve it: Modern heuristics.* Berlin, Germany: Springer–Verlag.

Min, H., Jayaraman, V., and Srivastava, R. (1998). Combined location-routing problems: A synthesis and future research directions. *European Journal of Operational Research* 108:1–15.

Murata, T., and Ishibuchi, H. (1994). Performance evaluation of genetic algorithms for flow shop scheduling problems. *Proceedings of 1st IEEE International Conference on Evolutionary Computation*, Orlando, FL, 812–817.

Nagurney, A. (1999). Economic equilibrium and financial networks. *Mathematical and Computer Modeling* 30:1–6.

——— (2000). *Sustainable transportation networks*. Cheltenham, UK: Edward Elgar Publishing.

——— (2006). On the relationship between supply chain and transportation network equilibria: A supernetwork equivalence with computations. *Transportation Research Part E* 42:293–316.

——— (2010). Supply chain network design under profit maximization and oligopolistic competition. *Transportation Research Part E* 46:281–294.

Nagurney, A., Cruz, J., Dong, J., and Zhang, D. (2005). Supply chain networks, electronic commerce, and supply side and demand side risk. *European Journal of Operational Research* 164:120–142.

Nagurney, A., Dong, J., and Zhang, D. (2002). A supply chain network equilibrium model. *Transportation Research Part E* 38:281–303.

Nagurney, A., and Toyasaki, F. (2005). Reverse supply chain management and electronic waste recycling: A multitiered network equilibrium framework for e-cycling. *Transportation Research Part E* 41:1–28.

Nagurney, A., and Yu, M. (2012). Sustainable fashion supply chain management under oligopolistic competition and brand differentiation. *International Journal of Production Economics* 135:532–540.

O'Kelly, M. (1987). A quadratic integer program for the location of interacting hub facilities. *European Journal of Operational Research* 32:393–404.

Outrata, J., Kocvara, M., and Zowe, J. (1998). *Nonsmooth approach to optimization problems with equilibrium constraints*. Dordrecht, The Netherlands: Kluwer Academic Publishers.

Patriksson, M. (1994). *The traffic assignment problem—Models and methods*. Utrecht, The Netherlands: VSP.

——— (2008). On the applicability and solution of bilevel optimization models in transportation science: A study on the existence, stability and computation of optimal solutions to stochastic mathematical programs with equilibrium constraints. *Transportation Research Part B* 42:843–860.

Poorzahedy, H., and Rouhani, O. (2007). Hybrid meta-heuristic algorithms for solving network design problem. *European Journal of Operational Research* 182:578–596.

Poorzahedy, H., and Turnquist, M. A. (1982). Approximate algorithms for the discrete network design problem. *Transportation Research Part B* 16:45–55.

Radcliffe, N. J., and Surry, P. D. (1994). Formal memetic algorithms. In *Evolutionary computing*, ed. Fogarty, T. Berlin, Germany: Springer-Verlag.

Resende, M. G. C., and Pinho de Sousa, J. (2004). *Metaheuristics: Computer decision making*. Dordrecht, The Netherlands: Kluwer Academic Publishers.

ReVelle, C. S., and Laporte, G. (1996). The plant location problem: New models and research prospects. *Operations Research* 44:864–874.

Rezapour, S., and Farahani, R. Z. (2010). Strategic design of competing centralized supply chain networks for markets with deterministic demand. *Advanced Engineering Software* 41:810–822.

Ribeiro, C., and Hansen, P. (2001). *Essays and surveys on metaheuristics*. Dordrecht, The Netherlands: Kluwer Academic Publishers.

Rochat, Y., and Taillard, E. D. (1995). Probabilistic diversification and intensification in local search for vehicle routing. *Journal of Heuristics* 1:147–167.

Sheffi, Y. (1985). *Urban transportation networks: Equilibrium analysis with mathematical programming methods*. Englewood Cliffs, NJ: Prentice Hall.

Southworth, F., and Peterson, B. E. (2000). Intermodal and international freight modeling. *Transportation Research Part C* 8:147–166.

Steenbrink, A. (1974). Transport network optimization in the Dutch integral transportation study. *Transportation Research Part B* 8:11–27.

Syswerda, G. (1989). Uniform crossover in genetic algorithms. In *Proceedings of Third International Conference on Genetic Algorithms*. Fairfax, VA: Morgan Kaufmann Publishers, Inc., 2–9.

Tavasszy, L. A. (1996). Modeling European freight transport flows. Unpublished doctoral dissertation, Delft University of Technology, Delft, The Netherlands.

Thomas, R. (1991). *Traffic assignment techniques*. Aldershot, UK: Avebury Technical.

Ukkusuri, S. V., and Patil, G. (2009). Multiperiod transportation network design under demand uncertainty. *Transportation Research Part B* 43:625–642.

Wardrop, J. G. (1952). Some theoretical aspects of road traffic research. *Proceedings of Institute of Civil Engineers II*, 325–378.

Yamada, T., Imai, K., Nakamura, T., and Taniguchi, E. (2011). A supply chain–transport supernetwork equilibrium model with the behavior of freight carriers. *Transportation Research Part E* 47:887–907.

Yamada, T., Russ, B. F., Castro, J., and Taniguchi, E. (2009). Designing multimodal freight transport networks: A heuristic approach and applications. *Transportation Science* 43:129–143.

Yang, G., Wang, Z., and Li, X. (2009). The optimization of the closed-loop supply chain network. *Transportation Research Part E* 45:16–28.

Yang, H., and Bell, M. G. H. (1998). Models and algorithms for road network design: A review and some new developments. *Transportation Review* 18:257–278.

Zhao, L., and Nagurney, A. (2008). A network equilibrium framework for Internet advertising: Models, qualitative analysis, and algorithm. *European Journal of Operational Research* 187:456–472.

CHAPTER 8

Vehicle Routing and Scheduling with Uncertainty

Ali Gul Qureshi

CONTENTS

8.1 Introduction	190
8.2 The Vehicle Routing and Scheduling Problem with Time Windows	190
8.3 The Dynamic Vehicle Routing and Scheduling Problem with Time Windows	193
8.3.1 Dynamic Customer	193
8.3.1.1 Degree of Dynamism	194
8.3.1.2 Representation of Dynamism	196
8.3.1.3 Diversion Issue and Waiting Policies	196
8.3.1.4 Rejection Policy	197
8.3.2 Dynamic Travel Time	197
8.3.3 Test Instances	199
8.3.4 Solution Approaches	200
8.3.4.1 Exact Optimization for the DVRPTW	201
8.3.4.2 Heuristics for the DVRPTW	203
8.4 The Stochastic Vehicle Routing Problem (SVRP)	206
8.4.1 Chance-Constrained Model	207
8.4.2 Recourse Model	209
8.4.3 Instances	210
8.4.4 Solution Approaches	211
8.4.4.1 Exact Approaches	211
8.4.4.2 Heuristics Approaches	214
8.5 The Stochastic Vehicle Routing Problem with Time Windows	215
8.6 The Stochastic and Dynamic Vehicle Routing Problem with Time Windows	217
8.6.1 Waiting Strategies	218

8.6.2 Instances 218
8.6.3 Solution Approaches 219
References 220

8.1 INTRODUCTION

After the December 2004 tsunami that devastated many coastal regions in some South Asian and Southeast Asian countries, and due to the logistics challenges faced in recent large-scale natural disasters such as an earthquake and a tsunami in Japan (March 2011) and earthquakes in Pakistan (October 2005) and China (May 2008), the importance of emergency logistics planning in Asian megacities has increased many fold. Although a vast body of operations research literature deals with the dynamic and stochastic nature of the business logistics, such literature about emergency logistics is limited. One of the main characteristics that differentiate emergency logistics from business logistics is the predominant presence of uncertainty in demand, supply, and transportation networks (i.e., travel times) (Sheu 2007). For example, actual demand of relief and evacuation is always revealed during the actual relief operation (dynamic demand)—even though a demand forecast is usually available beforehand based on some probabilistic models. Hence, a good practice in emergency logistics planning can be the use of dynamic demand models with some stochastic information on demand.

Many researchers have modeled emergency relief distribution as a dynamic mixed integer multicommodity network flow problem (e.g., see Haghani and Oh 1996; Ozdamar, Ekinci, and Kucukyazici 2004). It lies at the juncture of many dynamic and stochastic variants of the vehicle routing problem (VRP), such as the dynamic pickup and delivery problem, and the split delivery vehicle routing problem. Since its inception in 1959 (Dantzig and Ramser 1959), the VRP has been one of the most fertile areas of research in the operations research and transportation fields; many variants of the VRP have been introduced, each with an abundance of research. This chapter provides an introduction to two variants of the VRP model closely related to the emergency logistics: the dynamic vehicle routing and scheduling problem with time windows (DVRPTW) and the stochastic vehicle routing and scheduling problem (SVRP).

8.2 THE VEHICLE ROUTING AND SCHEDULING PROBLEM WITH TIME WINDOWS

The classical VRP consists of finding a set of minimum cost routes for k vehicles stationed at a central depot to cover all demands d_i of geographically located n customers with a constraint that the sum of demands along a route shall be less than the vehicle capacity q. Due to this *capacity constraint,* it is also called the capacitated vehicle routing problem (CVRP). The vehicle routing and scheduling problem with time windows results are due to the addition of the *time windows constraint* to the VRP definition. This constraint ensures the start of service at each customer i within its prespecified time window $[a_i, b_i]$, where a_i specifies the earliest possible service time and b_i is the desired latest possible service time at the concerned customer location.

The VRPTW is defined on a directed graph $G = (V, A)$, where the vertex set V includes the depot vertex 0 and set of customers $C = \{1, 2,..., n\}$. The set of identical vehicles with capacity q stationed at the depot is represented by K. With every vertex of V there is associated a demand d_i, with $d_0 = 0$, and a time window $[a_i, b_i]$ representing the earliest and the latest possible service start times. Cost c_{ij} as well as time t_{ij} are associated with each arc of the graph. The time t_{ij} includes the travel time on arc (i, j) and the service time at vertex i. The arc set A consists of all feasible arcs (i, j) satisfying the inequality $a_i + t_{ij} \le b_j$, $i, j \in V$. Furthermore, s_{jk} defines the service start time at a vertex $j \in C$ by a vehicle $k \in K$. Using the typical three index formulation commonly used in the related literature, the VRPTW can be mathematically described as Equations (8.1)–(8.9).

The model contains two decision variables: s_{jk} and x_{ijk}. If the customer j is served by a vehicle k, s_{jk} determines the service start time at customer j. The decision variable x_{ijk} represents whether arc (i, j) is used in the solution ($x_{ijk} = 1$) or not ($x_{ijk} = 0$). The objective function (8.1) minimizes the total cost that may include the fixed cost for vehicle utilization if it is added to the cost of all outgoing arcs from the depot (i.e., to c_{0i}) and the travel cost on the arcs with $x_{ijk} = 1$. Constraint (8.2) ensures that every customer is only serviced once, while constraint (8.3) is the capacity constraint that keeps the cumulative demand along a vehicle's route within its capacity. Constraints (8.4)–(8.6) are flow conservation constraints, specifying that every route shall start from and end at a depot; while on the route, if a vehicle travels to a customer location h, it must also travel from it. Constraint (8.7) is a time window constraint specifying that if a vehicle travels from i to j, service at j cannot start earlier than that at i. Here, M is a large constant. Constraint (8.8) represents the time windows that restrict the service start time at all vertices within their time windows $[a_i, b_i]$. Finally, constraint (8.9) ensures the integrality of the flow variables x_{ijk}. Note that the constraint represented by Equation (8.8) is the linearized form of the original nonlinear constraint given by Equation (8.10):

$$\min \sum_{k \in K} \sum_{(i,j) \in A} c_{ij} x_{ijk} \qquad (8.1)$$

subject to

$$\sum_{k \in K} \sum_{j \in v} x_{ijk} = 1 \quad \forall i \in C \qquad (8.2)$$

$$\sum_{i \in C} d_i \sum_{j \in V} x_{ijk} \leq q \quad \forall k \in K \qquad (8.3)$$

$$\sum_{j \in V} x_{0jk} = 1 \quad \forall k \in K \qquad (8.4)$$

$$\sum_{i \in V} x_{ihk} - \sum_{j \in V} x_{hjk} = 0 \quad \forall h \in C, \quad \forall k \in K \qquad (8.5)$$

$$\sum_{i \in V} x_{i0k} = 1 \quad \forall k \in K \qquad (8.6)$$

$$S_{ik} + t_{ij} - S_{jk} \leq (1 - x_{ijk}) M_{ijk} \quad \forall (i,j) \in A, \quad \forall k \in K \qquad (8.7)$$

$$a_i \leq s_{ik} \leq b_i \quad \forall i \in V, \quad \forall k \in K \qquad (8.8)$$

$$x_{ijk} \in \{0,1\} \quad \forall (i,j) \in A, \quad \forall k \in K \qquad (8.9)$$

$$x_{ijk}(s_{ik} + t_{ij} - s_{jk}) \leq 0 \quad \forall (i,j) \in A, \forall k \in K \qquad (8.10)$$

In the case of the soft time windows, the upper bound in the constraint (8.8) is relaxed allowing late arrivals and a late arrival penalty (C_l) (Equation 8.11) is added to the objective function (see Branchini, Armentano, and Løkketangen 2009; Larsen 2001; Gendreau et al. 1999). Similarly, a vehicle that arrives early has to wait till the opening of time windows (i.e., till a_i), while the earlier mentioned references consider this waiting without any cost. Taniguchi and Shimamoto (2004) and Fleischmann, Gnutzmann, and Sandvoss (2004) assigned an early arrival penalty (C_e) with this waiting and added the corresponding term (Equation 8.12) in the objective function (Equation 8.1). The arrival time of the vehicle k at customer j is represented by s'_{jk}, and the unit late arrival and early arrival penalties are shown by c_l and c_e respectively. The exponent α is mostly taken as unity; however, Fleischmann, Gnutzmann et al. (2004) have used $\alpha = 2$.

$$C_l = c_l (max(s_{jk} - b_j, 0)^\alpha), \text{ if } s_{jk} > b_j \qquad (8.11)$$

$$C_e = c_e \left(max(a_i - s'_{jk}, 0)^\alpha \right), \text{ if } s'_{jk} < a_j \qquad (8.12)$$

8.3 THE DYNAMIC VEHICLE ROUTING AND SCHEDULING PROBLEM WITH TIME WINDOWS

As mentioned in the introduction, the emergency relief distribution inherently belongs to the dynamic class of the VRP, where information is revealed during the execution of the relief operation. A routing system is defined as *dynamic* if complete or partial input information (such as number and location of customers or travel time on arcs) is not available to the decision maker at the start but is revealed during the scheduling horizon (day of operation), and if the decision maker *reacts* to this new information by evoking some sort of *reoptimization* mechanism (Psaraftis 1995). Thus, even a routing system would be termed as static if it incorporates a known variation of travel times (and corresponding travel costs) on arcs at the time of route optimization. Similarly, the settings used in Alvarenga, Silva, and Mateus (2005) also do not qualify to be called DVRPTW, where a few customers are inserted after the start of *solution algorithm,* rather than at the scheduling horizon, prior to the departure of vehicles from the depot. In the SVRP, the actual demand of a customer is unknown but it follows a known probability distribution; this problem will be tackled in detail later in the chapter. Many formulations and strategies adopted for the SVRP, which do not attempt any reoptimization when the actual demand of a customer is revealed, are also categorized under static systems.

Many inputs or elements of the VRPTW can impart dynamism in the problem such as customers' data, service times at customers, and travel times on the arcs. This section will highlight these sources of uncertainties and the related policies and solution approaches available in recent literature relating to the DVRPTW. There exists plenty of research on less constrained dynamic routing problems such as the dynamic traveling salesmen problem (DTSP) and dynamic vehicle routing problem (DVRP); interested readers can find an excellent review of these problems in Powell, Jaillet, and Odoni (1995).

8.3.1 Dynamic Customer

The case of dynamic customers or dynamic demand points is very clear in the context of emergency relief distribution after a natural disaster, such as earthquakes and tsunamis. New areas surface as the information and relief operation evolve in the affected area. In business logistics, dynamic customer characteristics are found in courier services, heating oil distribution, taxi services, dial-a-ride systems, etc.

8.3.1.1 Degree of Dynamism

Larsen, Madsen, and Solomon (2002) defined the degree of dynamism (*dod*) of a DVRP instance as the ratio of dynamic customers—immediate requests (n_{imm})—to the total number of customers (n) as given in Equation (8.13). The total number of customers includes the dynamic customers and the advance customers (n_{adv}) (i.e., the customers known prior to the start of the scheduling horizon). The advance customers are those who call in early in the morning before start of the operational day or those left unserved on the previous day.

$$dod = \frac{n_{imm}}{n} \quad (8.13)$$

Generally, the objective function of a static problem (*dod* = 0) is to minimize the routing cost; however, purely dynamic systems (*dod* = 100%), such as ambulance services, usually consider minimization of the response time as their objective function. Most of the business logistics problems fall in the category of partially dynamic systems (0 < *dod* < 100%) with their objective functions as a compromise between minimizations of the routing cost and the response time (Larsen et al. 2002). Figure 8.1 shows a classification framework suggested by Larsen et al. (2002) based on the concept of *dod* and the objective function. They classified a routing problem as a weakly dynamic system if its *dod* is less than 20%–30%, whereas a routing problem with *dod* greater than 80%–90% was classified as a strong dynamic system.

The concept of the *dod* ignores the arrival time of the dynamic customers. If an immediate request arrives early in the scheduling horizon, in the absence of time windows, decision makers have sufficient time to react as compared to a dynamic customer calling late in the scheduling horizon. To capture this effect, Larsen et al. (2002) suggested the use of effective degree of dynamism (*edod*) that exclusively considers the arrival time of each dynamic request. Let t_i^a represent the arrival time of an immediate request i and T show the scheduling horizon or depot's working hours ($T = b_0 - a_0$); the *edod* is defined as

$$edod = \frac{\sum_{i=1}^{n_{imm}} \left(\frac{t_i^a}{T}\right)}{n} \quad (8.14)$$

Note that, according to the *edod*, a purely dynamic system (*edod* = 1) receives all customer requests at time T, which is not possible in real-life instances unless the depot closing time is relaxed.

The *edod* considers the arrival times of the immediate requests in a routing environment without time windows. The presence of time

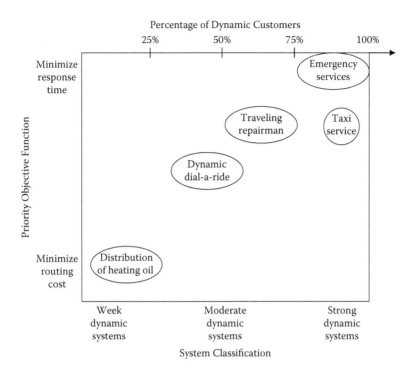

FIGURE 8.1 Classification framework for dynamic vehicle routing problems. (Source: Adapted from Larsen, A. et al., 2002, *Journal of Operations Research Society* 53:637–646.)

windows adds another complexity to a dynamic routing system as customer arrival time t_i^a and its time windows $[a_i, b_i]$ require the decision maker to react within a limited *reaction time* ($r_i = b_i - t_i^a$). Larsen et al. (2002) also defined the effective degree of dynamism in time windows constrained routing systems as the relation between the reaction time and the scheduling horizon as given in

$$edod = \frac{\sum_{i=1}^{n}\left(1-\frac{r_i}{T}\right)}{n} \qquad (8.15)$$

Larsen (2001) tested a DVRPTW with soft time windows by allowing lateness at some penalty cost and concluded that the average distance traveled and lateness increase with the increase in the *dod*. The results show that even 10% advance customers (*dod* = 90%) produce much less lateness as compared to the purely dynamic instances. Similar results are reported for the *edod* (see Larsen et al. 2002), though it was found that a higher value of *edod* means that the immediate request arrives later in the scheduling horizon providing sufficient time to serve advanced customers and resulting in decreased travel distance. The computation results given in Larsen (2001) show that the $edod_{tw}$ does not provide any better insight into the solutions properties of the DVRPTW compared to the simple *dod*.

8.3.1.2 Representation of Dynamism

The complete scheduling horizon (i.e., 0 to T) is usually divided into many slots, which are marked with occurrences of events. An event can be passage of a priori fixed time duration (say, ΔT) resulting in uniform division of the scheduling horizon (e.g., see Chen and Xu 2006) or it is based on arrival of a single or a batch of dynamic customers that provides nonuniform division of the scheduling horizon. Service completion at a customer and vehicle departure can also be considered as events. A term, *rolling horizon,* is used to describe the unfolding of different scenarios of a dynamic routing instance. Each scenario consists of a static instance; those customers serviced during the previous time slot are neglected in forming the next scenario while the immediate requests received during the previous time slot are added. The position of vehicles and their residual capacities are also included in the definition of the new scenario.

8.3.1.3 Diversion Issue and Waiting Policies

Ichoua, Gendreau, and Potvin (2000) introduced the concept of *diversion* in the DVRPTW with dynamic customers, which is also adopted in Chen and Xu (2006). In this policy, an en route vehicle with an assigned destination in the previous scenario can be diverted to a different customer in the current scenario. While testing this policy on a DVRPTW with soft time windows, they reported reduction in traveled distance and lateness against the typically followed policy of no diversion, which has been followed in Taniguchi and Shimamoto (2004) and Gendreau et al. (1999). At the unfolding of a new scenario, the location of a vehicle is forecast based on its current location, speed, and allocated time for reoptimization. Under the no-diversion policy, if a vehicle leaves for the next customer j immediately after finishing service at the current customer i, it is committed to the customer j even if it arrives earlier and waits there; that is, it cannot leave customer j without serving the customer. If a dynamic customer appears in the vicinity of customer i requesting immediate servicing, this situation would lead to an underoptimized solution. Therefore, in dynamic routing systems, waiting is considered better at the current/serviced customer; this strategy is also called the *least commit strategy.* Mitrovic-Minic and Laporte (2004) used the term *wait-first* for this waiting strategy while describing some other waiting strategies for the dynamic pickup and delivery problem with time windows (DPDPTW). Branchini et al. (2009) defined *positioning* as the case of a vehicle waiting after serving all its customers, just before leaving for the depot, at a prespecified idling point that corresponds to a location with a high expectancy of new customer arrival. They concluded that compared to the case when these strategies are not used, the combination of diversion, wait-first, and positioning provides much better results in terms of less distance traveled, fewer late arrivals, a high percentage of customer coverage, and fewer rejections.

8.3.1.4 Rejection Policy

Another issue with the modeling of the DVRPTW with dynamic customers is the authority of the decision maker to reject a particular immediate request. If rejection is not allowed, the objective function is usually based on minimizing the total operational cost (travel distance, etc.) (e.g., Chen and Xu 2006; Taniguchi and Shimamoto 2004). Reducing the number of unserviced/rejected customers is usually a comparative measure if the rejection is allowed (see Ichoua et al. 2000; Gendreau et al. 1999).

8.3.2 Dynamic Travel Time

Soon after a disaster, the uncertainty related to the availability/usability of the transportation network is very high. Some of the links available in some time interval may be unavailable in another and vice versa. This leads to the situation of dynamic travel times in emergency relief distribution, whereas the day-to-day business logistics experiences dynamic travel times due to ever changing traffic conditions and unexpected incidents. With advances in the technology, it has become easier and relatively economical to gather large amounts of real-time traffic data, such as travel times. This has stimulated the research in routing systems based on the fluctuating or variable travel times (for example, see Fleischmann, Gietz et al. 2004; Ichoua, Gendreau, and Potvin 2003). As discussed earlier, however, the VRPTW with time-dependent travel time is classified as a static problem (Psarafits 1995). While fairly abundant research considers the case of dynamic customers, the research concerning the DVRPTW with dynamic travel times is scant and we could spot only the work of Taniguchi and Shimamoto (2004), whereas Fleischmann, Gnutzmann et al. (2004) have presented a DPDPTW with dynamic customers and dynamic travel time. This may be due to the fact that successful implementation of a routing system with dynamic travel times is a very difficult task that requires integration of data from various sources such as city traffic units and logistics firms (Taniguchi et al. 2001).

Every arc—say, (i, j)—in the complete graph G represents a path between two customers that may contain many links of the underlying road network. An increase in the travel time of an arc (i, j) may occur due to increase in the travel time on a link at the end of the corresponding path. If a vehicle is traveling from i to j at the occurrence of an event that requires reoptimization based on the updated travel times, the diversion policy can be very effective. However, both Fleischmann, Gnutzmann et al. (2004) and Taniguchi and Shimamoto (2004) have not considered the diversion or the customer rejection policy.

One of the important issues with modeling of the DVRPTW with dynamic travel time is the source of the travel time data and their management. The dynamic travel times can be obtained via traffic simulations. For example, Taniguchi and Shimamoto (2004) have used a macroscopic traffic simulator to provide updated travel time for a test network after each minute; however, it was used in reoptimization whenever a vehicle completed service at a customer. The other obvious source is, of course, the intelligent transport systems (ITS) that continuously monitor the traffic situation on urban road networks and report the travel times and speeds on various links such as the vehicle information and communication system (VICS) in Japan. Fleischmann, Gnutzmann et al. (2004) used the data from such a system implemented in Berlin, Germany, and named LISB that provides the travel time data on links for every 5-minute slot.

Use of a real-life data source, however, poses some additional problems such as transformation of the actual road network (and corresponding link travel times) to a complete graph G (and corresponding arc travel times) used in the DVRPTW. Some critical questions are whether to neglect small and unusable (due to traffic restrictions) links or not and how to react to the changing travel times on road network links. For example, Fleischmann, Gnutzmann et al. (2004) used Dijkstra's algorithm to find shortest paths between customers and depots based on the time-dependent travel times suggested by Fleischmann, Gietz, and Gnutzmann (2004). In the dynamic settings, these paths were monitored for any change in the travel time of a link constituting such a path. The second question relates to the consideration of local or global impact of the change in a link's travel time—that is, the propagation of this change in the road network as a change in other links. Fleischmann, Gnutzmann et al. (2004) used only local impact of a minor travel time change of a link by modifying the length of corresponding shortest paths using such a link, and they considered the global impact only when a major change in travel time occurred by considering the related changes on the preceding links. However, the shortest paths (i.e., the sequence of links) were subjected to change only if a link on a particular shortest path was blocked completely.

The traffic simulation method, on the other hand, can be used to sparse the underlying network by eliminating smaller links at the time of network coding and, also, by just considering the related area of the network; however, it does not guarantee the exact real-life behavior. While macroscopic simulation is fairly fast, the recent advancement in computational power will tempt researchers to use microscopic simulation for better replication of real-life effects of dynamic travel times.

Similar to the dynamic customers' case, the representation of the dynamism in the dynamic travel time case is based on the rolling horizon. In a DVRPTW with only dynamic travel times, an event can be completion of service at a customer (Taniguchi and Shimamoto 2004) or a considerable change in travel time (Fleischmann, Gnutzmann et al. 2004). If dynamic customers are also considered along with dynamic travel time, a combination of these events and events described in Section 8.3.1.2 can represent a change in the scenarios of the rolling horizon (see Fleischmann, Gnutzmann et al. 2004).

8.3.3 Test Instances

No benchmark data set is available for the DVRPTW, which may be due to the randomness and uncertainty attached to the arrival time of dynamic customers and dynamic travel times. However for the DVRPTW, few researchers have modified the well known Solomon's benchmark instances (Solomon 1987) commonly used to evaluate the performance of models and algorithms for the VRPTW. In case of dynamic customers, two issues need to be decided in the modification of a Solomon's benchmark instance: namely, the proportion of immediate requests (or the *dod*) and their arrival time (or call-in time). Let T' represent the ratio of the scheduling horizon to the depot operation time in Solomon's benchmark instances given by Equation (8.16). Chen and Xu (2006) considered a uniform distribution for the arrival time in the interval shown in Equation (8.17), where τ shows the computation time allocated for the optimization after each event. Customers with a negative t_i^a value were regarded as the advance customers and the rest as the dynamic customers. Gendreau et al. (1999) and Ichoua et al. (2000) consider half of the customers in the Solomon's benchmark instances as immediate requests, with their arrival time being a random value generated within the interval shown in Equation (8.18) where s_{i-1}^d represents the departure time from the predecessor of the customer i in a best solution of the static version of the corresponding Solomon's benchmark instance. Ichoua et al. (2003) also considered test instances with *dod* = 50% and 75%, using the same settings for request arrival time as

Gendreau et al. (1999). They considered three different arc categories and variable travel time (three slots) for each category:

$$T' = \frac{T}{b_0 - a_0} \tag{8.16}$$

$$t_i^a \in \left[\frac{1}{2}\min\{T'a_i, Tb_i - t_{0i} - \Delta T - \tau\}, \min\{T'a_i, Tb_i - t_{0i} - \Delta T - \tau\}\right] \tag{8.17}$$

$$t_i^a \in \left[0, T' \min\{a_i, s_{i-1}^d\}\right] \tag{8.18}$$

Fleischmann, Gnutzmann et al. (2004) presented their computational results of the DVRPTW with dynamic customers and dynamic travel time on a test instance completely based on the real-life data; however, such cases are rare and mostly artificially generated instances have been used in the DVRPTW related research. Euclidean distance-based instances are developed by generating the customers' locations (coordinates) in a bounded area, using a random uniform distribution (Larsen 2001), or by favoring certain portions to mimic a real-life situation (Branchini et al. 2009). Poisson distribution has been adopted in most of the DVRPTW-related literature to simulate the arrival times of the dynamic customers to generate instances for a particular value of *dod*. Branchini et al. (2009) have used a time-based intensity λ_t to represent higher request arrival rates during the earlier part of the scheduling horizon. Larsen (2001) has considered a fixed service time, whereas, in real life, it can be different for different customers. Therefore, for a better representation, it can be considered a random value sampled from some distribution such as a normal distribution (Branchini et al. 2009). Larsen et al. (2002) have considered a log-normal distribution for service times, which provides closer real-life variations of the service time. Similarly, customers' time windows may be based on some probability distribution (Branchini et al. 2009) or using some fixed rules based on the request arrival time.

8.3.4 Solution Approaches

The dynamics of the DVRPTW problems require a quick response to the changing inputs (travel time or customers). This situation has favored heuristics approaches as these are considered faster compared to the exact-solution approaches, which traditionally have been computationally intensive. A direct comparison of various heuristics solutions suggested for the DVRPTW is also not possible as their performances have been typically accessed on a set of varied data sets. This section discusses some of the typical heuristics techniques common in the DVRPTW-related research along with an exact optimization approach for the dynamic travel time case.

8.3.4.1 Exact Optimization for the DVRPTW

Using a rolling horizon and event-based representation of the dynamic vehicle routing and scheduling with soft time windows with dynamic travel times (DVRPSTW), Qureshi, Taniguchi, and Yamada (2012a) proposed a column generation-based exact optimization algorithm for the DVRPSTW. Their DVRPSTW settings only considered dynamic travel time; all other inputs such as customers' locations and demands were assumed to be known in advance. Thus, at the change of time slot, the locations of en route vehicles were forecast and the underlying graph was updated to incorporate the dynamic changes in travel times—in this case, disruptions due to traffic incidents. In the updated graph, vehicles were available at the central depot as well as at some already served customers, which required consideration of a multidepot dynamic vehicle routing and scheduling problem with soft time windows (MD-VRPSTW), as each node with a vehicle resource should be considered as a virtual depot. The column generation-based approach developed for the static VRPSTW by Qureshi, Taniguchi, and Yamada (2009) was extended to solve the D-VRPSTW by formulating a separate elementary shortest path problem with resource constraint and late arrival penalty (ESPPRCLAP) subproblem for each of the virtual depots. At a column generation iteration, the set partitioning master problem receives columns (routes/paths of negative reduced costs) from all subproblems and then it optimizes the complete problem by selecting the best set of routes covering the demands of all the customers at hand. The dual variables' values (*prices*) obtained as a by-product are then used to define new reduced-cost matrices for each of the subproblems to generate new promising paths/columns, and the whole process is repeated till the subproblems fail to provide a negative reduced-cost column. At this stage, if the solution of the master problem linear program (LP) is not an integer, a branch and bound tree is explored. Therefore, at every route revision epoch, the column generation algorithm is embedded in a branch and price algorithm. In order to track and keep the total number of vehicles in the system, a branching scheme has also been introduced that first ensures an integer number of vehicles departing from each depot and then aims at integer flow variables.

Later, in Qureshi et al. (2012b), a case study of the DVRPSTW was presented deriving the dynamic data from a microsimulation developed in the VISSIM ((Verkehr in Städten—Simulations)) software in order to mimic the actual traffic situation under normal and incident cases. Figure 8.2 gives the flow chart of the column generation algorithm

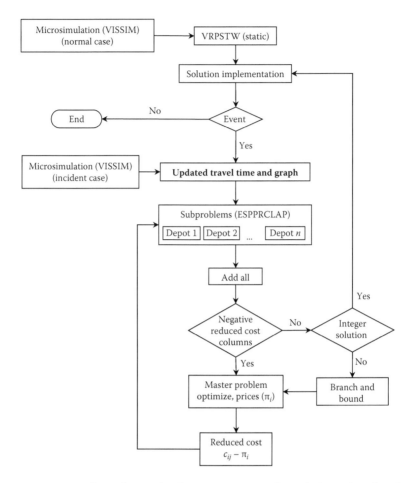

FIGURE 8.2 Flow chart of column generation-based algorithm for the DVRPSTW. (Source: Qureshi, A. G. et al., 2012b, *Procedia—Social and Behavioral Sciences* 39:205–216; http://dx.doi.org/10.1016/j.sbspro.2012.03.102)

proposed by Qureshi et al. (2012a) along with addition of the microsimulation step. The microsimulation of their test network contains two types of roads: major roads consisting of two lanes in each direction and minor roads of one lane in each direction. It covers approximately an area of 8 × 8 km² with five different vehicle classes: passenger car, bus, heavy vehicles (trucks), large motorcycle (with same maximum speed of 50 km/h as for cars, buses, and trucks), and light motorcycles (with maximum speed of 30 km/h). To obtain a test instance for the DVRPSTW, the time windows and the demand data of the first twenty-four customers of the Solomon's benchmark instance R101-100 were taken; their locations were redefined on the test network. The travel time data in both normal and dynamic conditions (traffic incident case) were obtained by running five-hour microsimulations in which the first hour was considered as the warm-up time.

In Qureshi et al. (2012a) a simple strategy was considered for the static VRPSTW that calls back the vehicle if it becomes certain that the next customer will become infeasible (due to lateness beyond soft time windows under the dynamic change of travel time); instead, a new vehicle is dispatched to cover the remaining route of the called-back vehicle. On the other hand, in the DVRPSTW case the system is reoptimized, which results in significant cost savings as well as less delay penalties.

8.3.4.2 Heuristics for the DVRPTW

Low computation time requirements and easy implementation of the local optimization-based heuristics such as the nearest neighborhood (NN), insertion heuristics (IH), and assignment heuristics (AS) make them an attractive option as solution techniques of the DVRPTW. These heuristics have also been used to evaluate the performance of some sophisticated heuristics such as metaheuristics and optimization-based heuristics. Larsen (2001) has adopted the push forward insertion heuristics (PFIH) developed for the static VRPTW by Solomon (1987). The solutions obtained from IH are often improved using some local search strategies such as 2-opt and OR-opt. Fleischmann, Gnutzmann et al. (2004) used OR-opt with PFIH and compared its combinations with AS for an *open ended* DPDPTW in which constraint (8.6) is relaxed. If the return to the depot is not restricted at the end of the route, greedy optimization usually performs better, as was the case with AS in Fleischmann, Gnutzmann et al. (2004).

Branchini et al. (2009) have observed that under some conditions, the capacity of the remaining vehicles available at the depot is underutilized if the advance customers are routed using the minimum required vehicles. To offset such situations, they proposed a two-stage heuristics approach: At the start of the operation a construction heuristics is used to spread the available vehicles and then a *granular* local search is used for the optimization of the dynamic requests. In the construction heuristics, *seed* customers equal to the number of available vehicles are chosen based on the concept of separation, which is based on intercustomer distance and their time windows. In the granular local search, the *granularity threshold* is defined—for example, a multiple of the average cost of an arc in a known solution. Then, a neighborhood (or candidate list of customers) is constructed consisting of arcs whose cost is less than the granularity threshold and some important arcs such as connecting it with depot. Thus, a sparse graph $G'(V, A')$, where A' consists of the neighborhood of all customers, is used in order to reduce the computation time for the formed local search. In the second stage, Branchini et al. (2009) have used an adaptive granular local search in which the cardinality of the set A' is increased or decreased with the decrease or increase in the arrival rate of immediate requests respectively.

Metaheuristics such as tabu search (TS) and genetic algorithms (GAs) have been successfully used for the static VRPTW. In the DVRPTW case, their relatively higher computational requirement poses some problems, especially when rejection policy is adopted. Nonetheless, an appropriate implementation can yield a much better optimization compared to the local optimization-based heuristics, such as in the case of a parallel implementation of an adaptive TS by Gendreau et al. (1999). In their algorithm, an adaptive memory (AM) is maintained that stores a fixed number of high quality solutions generated during the heuristics, by replacing the worst with better, if required. In the parallel implementation, I different initial solutions are formed using routes of the solutions present in the AM submitted to R decomposition processes, each decomposing these initial solutions into D subproblems, each of which is subjected to a separate TS on P separate processors connect in parallel to a master processor. The TS consists of the CROSS operator that exchanges a chain of customers (maximum of seven customers) between two routes. The obtained routes in each subproblem are then combined to reconstruct a solution and, after E cycles of such decompositions/reconstructions, the solution is sent for possible insertion in the AM. To regulate the time dedicated for the optimization, the E value is dynamically adjusted based on the average number of calls to the adaptive memory (with respect to R) in the last time slot. At an event of a new immediate request, all the search threads are interrupted and the corresponding solutions are sent for the inclusion in the AM. The new request is inserted at a feasible location (with respect to the hard time windows at the depot). If no feasible insertion is obtained in any of the solutions in the AM, the request is rejected; otherwise, only the feasible solutions are kept in the AM. The best feasible solution in the AM is reoptimized using CROSS operator and *best first* strategy, and then again new initial solutions are formed and the previously mentioned process is repeated. The best solution in the AM is supposed to be implemented; in the event of service completion to a customer by one of the vehicles of this solution, its next customer is fixed as its current destination. The remaining solutions present in the AM are also modified so that the current destination is the same for that vehicle in all solutions. The optimization continues based on the solutions available in the AM until the next event; therefore, the best solution that is being implemented may get changed.

To facilitate the implementation of diversion policy, instead of fixing the current destination, Ichoua et al. (2000) utilized the concept of current location. At the arrival time of a new immediate request (t_i^a), the location of a vehicle is forecasting at time $(t_i^a + \tau)$. The value of τ is regulated based on the moving average of the inter-request arrival time in the previous time slots or on the average time per request during a fraction of the scheduling horizon; the latter gave better results in computational experiments of Ichoua et al. (2000). During the interval $[t_i^a, \tau]$, the time is considered as frozen, and the plan of the previous time slot is executed. Meanwhile, a new arrival request in $[t_i^a, \tau]$ is handled using a copy of AM, which is kept updated with new solutions found in $[t_i^a, \tau]$. In order to incorporate the variable travel time consideration in the adaptive TS of Gendreau et al. (1999), Ichoua et al. (2003) used approximations (or proxies) to obtain the objective function values. Under the wait-first policy, the vehicle has to wait at i; a reverse recursive function was adopted to determine the latest possible departure time from customer i to j.

Taniguchi and Shimamoto (2004) have proposed a GA to solve a DVRPTW with dynamic travel times. At every event of service completion, the GA attempts to solve a static VRPTW instance based on the updated travel time information provided by a traffic simulator. The population is composed of 300 individuals, each representing a complete feasible solution with vehicle departure times and customer order. The population evolution is based on an ordered crossover, and an inversion mutation operator is used to broaden the search.

Some efficient heuristics approaches based on the exact optimizations framework have also been developed, such as dynamic column (DYCOL) generation heuristics presented by Chen and Xu (2006). Following the Dantzig–Wolfe decomposition concept, the DVRPTW with dynamic customers is formulated as a set partitioning master problem, whereas a local search (LS) heuristics is used to solve the subproblem that provides columns (single vehicle routes) of negative reduced cost based on the dual variables (π) generated in the master problem.

Chen and Xu (2006) divided the scheduling horizon uniformly into time slots bounded by a *decision epoch* T_i; the *implementation epoch* occurs at time $T_i + \tau$. At time T_{i-1}, the $(i-1)$th static problem is solved using the MIP (mixed integer programming) solver of CPLEX based on the available columns and the solution is implemented during the interval $[T_{i-1} + \tau, T_i + \tau]$. During the time interval $[T_{i-1} + \tau, T_i]$, DYCOL generates columns to be used for the MIP solution of the ith static problem, considering the vehicles' locations and their corresponding residual capacities at time $T_i + \tau$. First, the master problem LP that consists of already generated columns up to $T_{i-1} + \tau$ is modified by deleting the customers (and corresponding rows) to be serviced during $[T_{i-1} + \tau, T_i + \tau]$. Single customer columns (0-i-0) for every new immediate request since T_{i-1} are then added to the set of these modified columns, simultaneously adding additional rows for the new immediate request. The master problem

LP is optimized using CPLEX and the corresponding dual variables' values are obtained, which are used to define the reduced cost. A local search procedure is then activated that generates a fixed number of new columns of negative reduced costs in the neighborhood (defined by 2-exchange and OR-exchange) of each of the columns (routes) in the *basis* of the LP. These new columns are added to augment the master problem LP along with the new single-customer columns for the immediate request arrived in the last column generation iteration. This iterative column generation process continues till T_j, when the MIP solver is called to generate the solution to be implemented in the next time slot.

Before the start of the scheduling horizon, for the advance customers, DYCOL is initialized with single-customer (0-*i*-0) type columns. The previously mentioned column generation process is used to generate the columns, with the difference that no new immediate request arrives and the column generation continues until the local search heuristics-based subproblem fails to provide a new column of negative reduced cost.

8.4 THE STOCHASTIC VEHICLE ROUTING PROBLEM (SVRP)

The quantum of research about the stochastic vehicle routing and scheduling problem with time windows (SVRPTW) is very limited; hence, before discussing the SVRPTW, the stochastic version of the classical VRP problem is discussed here, which provides the foundations to the SVRPTW. In the SVRP, some of the inputs (such as customers' demand or travel time) are random variables. Although the actual values of such inputs are unknown and are revealed at the time of service, a probability distribution is available for these random variables. Using historic data, a probability distribution can be made available for the amount of relief required in case a disaster happens; the same is the case with the travel time distribution. Therefore, the predisaster scenario (i.e., at the preparedness stage) of the relief distribution can be typically modeled using the SVRP. In business logistics terms, stochasticity occurs in travel times due to varied traffic conditions, whereas deliveries to vending machines and ATM machines, pickup of cash from banks at the end of day, beer distributions to shops, and heating oil distribution to households are typical cases of the SVRP with stochastic customer demand.

Similarly to the case of the DVRPTW, uncertainties in customer demand have received more attention and the literature considering the vehicle routing problem with stochastic travel times demand (SVRP-T) is rather scant (Laport, Louveaux, and Mercure 1992). In the vehicle routing problem with stochastic demand (SVRP-D), the customer demand is no longer a fixed value (as defined in Section 8.2); rather, it is a random variable ξ_i with a mean of μ_i and a standard deviation of σ_i; each realization of ξ_i is termed a "state of world" (Taniguchi et al. 2001). Customers' stochastic demands ξ_i can be independent random variables, as mostly considered (see Dror and Trudeau 1986; Christiansen and Lysgaard 2007; Tan, Cheong, and Goh 2007), or may be correlated to each other (Stewart and Golden 1983). Sometimes, ξ_i is considered as discrete, having l different mutually exclusive values ξ_i^l, each with a probability of p_i^l (see Yang, Mathur, and Ballou 2000). The actual demand of a customer is revealed when the delivery/pickup vehicle arrives at the customer. In the vehicle routing problem with stochastic customers and demand (SVRP-CD), each customer i is present with a probability p_i and with a stochastic demand ξ_i as described earlier (Gendreau, Laporte, and Seguin 1995, 1996a). The presence of a customer is revealed when the service at the predecessor node begins while its demand is made available when a vehicle reaches it. An excellent survey of the theoretical issues of SVRP and related problems can be found in Powell et al. (1995); Gendreau et al. (1996b) provided a comprehensive and concise list of various distributions used for modeling the stochastic demand, such as normal, Poisson, binomial, etc.

Unlike the DVRPTW, the routes to be followed by the delivery/pickup vehicles are defined prior to the start of the scheduling horizon, giving the *a priori solution* (Bertsimas, Jaillet, and Odoni 1990). During the execution of these a priori routes, a *route failure* may occur when, for example, on arrival to a customer its actual demand comes out to be more than its expected demand (considered earlier) and the current residual vehicle capacity is insufficient to meet this demand. If the probability of such route failure is restricted to be within a prespecified limit (say, β), the resultant SVRP model is termed a *chance-constrained model*. Even though the chance of route failure is constrained, if it happens, a corrective measure called the *recourse action* (or *recourse policy*), has to be followed such as returning of the concerned vehicle to the depot, unloading its content to regain the capacity, and then continuing with the same a priori route (static problem) or with a different route (dynamic problem). The chance constraint model fails to take into account the cost of the recourse action; however, the *recourse model* of the SVRP assigns a penalty cost to this recourse and takes that into the objective function.

8.4.1 Chance-Constrained Model

The SVRP also contains constraints similar to those in the VRPTW formulation (shown in Section 8.2), such as constraints (8.2), (8.4)–(8.6), and (8.9). Due to stochastic demand in the SVRP-D and SVRP-CD, the capacity constraint is handled and represented in a different manner as compared to the VRPTW. Furthermore, in the chance-constrained formulation of the SVRP (Equations 8.19 and 8.20), the objective function also has the same form as Equation (8.1). However, for the sake of simplicity, it has been given again here:

$$\min \sum_{k \in K} \sum_{(i,j) \in A} c_{ij} x_{ijk} \tag{8.19}$$

subject to

$$\Pr\left[\sum_{i \in C} \xi_i \sum_{j \in V} x_{ijk} \leq q\right] \geq 1 - \beta \quad \forall k \in K \tag{8.20}$$

and constraints (8.2), (8.4) – (8.6) and (8.9).

As discussed earlier, the chance-constrained model restricts the probability of route failure to be less than a limit β, as shown in Equation (8.20). If the customers' demands (ξ_i) are independent random variables and the distribution of $\left(\sum_{i \in C} \xi_i \sum_{(i,j) \in A} x_{ijk} - M_k\right)/S_k$ is the same as that of ($\xi_i - \mu_i$)/σ_i, where $M_k = \sum_{i \in C} \mu_i \sum_{(i,j) \in A} x_{ijk}$ and $S_k = \left(\sum_{i \in C} \sigma_i^2 \sum_{(i,j) \in A} x_{ijk}^2\right)^{1/2}$ are the mean and standard deviation of the demands on a route k. Furthermore, if there exist some constants γ and θ, which satisfy Equations (8.21) and (8.22), Stewart and Golden (1983) proved that the chance-constrained SVRP model can be reduced to a deterministic VRP model by replacing constraint (8.20) with constraint (8.23), where the modified vehicle capacity $\bar{q}$ is given by Equation (8.24):

$$\alpha_i^2 = \gamma \mu_i \tag{8.21}$$

$$\Pr\left[\left(\sum_{i \in C} \xi_i \sum_{(i,j) \in A} x_{ijk} - M_k\right)\Big/S_k \leq \theta\right] = 1 - \beta \tag{8.22}$$

$$M_k \leq \bar{q} \tag{8.23}$$

$$\bar{q} = \left(2q + \theta^2 \gamma - \sqrt{\theta^4 \gamma^2 + 4q\theta^2 \gamma}\right)/2 \tag{8.24}$$

For a vehicle route $k = \left(i_{k_0} = 0, i_{k_1}, \ldots, i_{k_u}, i_{k_0}\right)$ of an SVRP with stochastic travel time and service time, Laporte et al. (1992) suggested a chance-constraint model by replacing constraint (8.20) with constraint

(8.25)—limiting the probability of the duration of a route exceeding a fixed value B to β. The service time of a customer i is represented by φ_i:

$$\Pr\left[\sum_{l=0}^{u}(t_{k_l k_{l+1}} + \phi_{k_l}) \leq B\right] \geq 1-\beta \quad \forall k \in K \quad (8.25)$$

8.4.2 Recourse Model

Although the chance constraint model of the SVRP is simple and easier to solve, especially when it can be reduced to a deterministic VRP model, the fact that it does not capture the cost of a recourse action to bring an infeasible solution (allowed with a probability of β) to feasibility reduces its effectiveness. Therefore, stochastic programming with recourse is most widely used to model the SVRP, which penalizes the route failure considering its cost, and instead optimizes the overall expected cost. Several recourse models for the SVRP can be found in the literature; a few of those and their variations are given here. Stewart and Golden (1983) suggested two very basic recourse models. In both models, the constraint (8.20) is Lagrangianly taken to the objective function; the first model (Equation 8.26) considers a fixed penalty ϕ_k, whereas the second model (Equation 8.27) considers the actual amount by which the constraint is violated, along with a unit penalty λ_k:

$$[\text{RC1}]\min \sum_{k \in K}\sum_{(i,j) \in A} c_{ij} x_{ijk} + \sum_{k \in K} \phi_k \Pr\left[\sum_{i \in C}\sum_{j \in V} \xi_i x_{ijk} > q\right] \quad (8.26)$$

subject to constraints (8.2), (8.4) – (8.6) and (8.9).

$$[\text{RC2}]\min \sum_{k \in K}\sum_{(i,j) \in A} c_{ij} x_{ijk} + \sum_{k \in K} \lambda_k E[l_k]$$

subject to constraints (8.2), (8.4) – (8.6) and (8.9).
where

$$E[l_k] = \sum_{l_k > 0} l_k \Pr\left[\sum_{i \in C}\sum_{j \in V} \xi_i x_{ijk} - q = l_k\right], \quad \forall k \in K \quad (8.27)$$

Many recourse policies for the SVRP have been suggested in the literature, some of which are provided here:

a. At the point of route failure, the simplest of all recourse policy is to service all the remaining customers by assigning a dedicated vehicle such as in Dror and Trudeau (1986).
b. In case of split delivery (by the same vehicle), it is also possible that the vehicle unloads the residual capacity at the point of route failure, returns to the depot to regain the capacity, goes back to the point of route failure, and continues the a priori route in the same order (Tan et al. 2007; Gendreau et al. 1995, 1996a).
c. Dror, Laporte, and Trudeau (1989) suggested a recourse policy that is considered as reoptimization and thus categorizes the corresponding SVRP as a dynamic problem (Psaraftis 1995). According to this policy, the demand of a customer is fully known at the arrival of the vehicle, but, instead of following recourse policy (b), keeping the knowledge about the customer's demand, the vehicle may proceed to another location which can be a different customer or the depot (for replenishment).
d. Instead of opting for the recourse only at the point of route failure, Yang et al. (2000) have utilized a restocking policy that enables the option of preemptive replenishment of capacity before the actual route failure.

8.4.3 Instances

Similarly to the DVRPTW, a set of standard SVRP instances does not exist. Stewart and Golden (1983) have adopted the VRP instances by Christofides and Eilon (1969) by considering the actual demand as mean demand and assigning a standard deviation as well. Dror and Trudeau (1986) gave the data of one such instance, which has been used by Tan et al. (2007) as a benchmark instance along with similarly modified Solomon's benchmark instances. To simplify further, Christiansen and Lysgaard (2007) assigned a Poisson distribution (with $\mu = \sigma^2$) to the demands in the VRP instances by Christofides and Eilon (1969).

In the self-generated test instances, customers' locations can be treated similarly to the case of the DVRPTW (Section 8.3). Various probability distributions can be used to simulate the stochastic nature of the demands, such as Poisson distribution (Laporte, Louveaux, and Mercure 2002), uniform discrete distribution (Gendreau et al. 1996a, 1995), discrete triangular distribution (Yang et al. 2000), etc.

8.4.4 Solution Approaches

8.4.4.1 Exact Approaches

Some efficient exact solutions approaches are available for the SVRP case—for example, the branch and cut based L-shaped method presented by Gendreau et al. (1995) for the SVRP-CD. Considering x as the a priori solution with cost cx, they modeled the objective function of the recourse model as Equation (8.28), where $Q^-(x)$ shows the negative penalty (cost savings) due to customer skipping with $(p_i = 0)$ and $Q^+(x)$ as the recourse penalty due to route failure. If a lower bound for expected cost of the second-stage solution $(\tilde{c}_{ij})$ (Equation 8.29) can be defined for an arc (i, j) belonging to the first-stage solution, the recourse model can be re-modeled as Equation (8.30), where $\tilde{Q}(x)$ is a non-negative penalty. Considering y as the lower bound on $\tilde{Q}(x)$, the L-shaped method solves the SVRP with the objective function given in Equation (8.31), neglecting the subtour elimination constraints. Initially, y starts at 0 and gradually increases using *optimality cuts*. These cuts, along with the subtour elimination, are added in a branch and bound algorithm in Gendreau et al. (1995):

$$\min_{x}\{cx - Q^-(x) + Q^+(x)\} \qquad (8.28)$$

$$\tilde{c}_{ij} = \begin{cases} p_j c_{ij}, i = 1, \forall j \in V \\ p_i p_j c_{ij} + \dfrac{1}{2}\left\{p_i(1-p_j)\min_{h \neq i,j} c_{ih} + p_j(1-p_i)\min_{h \neq i,j} c_{hj}, \forall i > 1, j \in V\right\} \end{cases} \qquad (8.29)$$

$$\min_{x}\{\tilde{c}x - \tilde{Q}(x)\} \qquad (8.30)$$

$$\min_{x,y}\{\tilde{c}x - y\} \qquad (8.31)$$

Let $T^k(x^v)$ (Equation 8.32) show the expected cost of a route $k = (i_{k_0} = 0, i_{k_1}, \ldots, i_{k_u}, i_{k_0})$. The expected solution cost $(T(x^v))$ with a feasible solution vector (x^v, y^v) of the L-shaped stochastic program (Equation 8.31) is given by Equation (8.36). Here, $\gamma_i^k(g)$ represents the non-negative penalty from i to depot, given that the vehicle k arrives at i with a residual capacity g. An optimality cut (Equation 8.37) is generated and added to the SVRP formulation, where $E^v = \{(i, j) | i, j > 1 \text{ and } x^v_{ij} = 1\}$, whenever the corresponding expected penalty $\tilde{Q}(x^v)$ (Equation 8.38) is greater than y^v:

$$T^k(x^v) = \sum_{i=k_1}^{k_u}\sum_{j=i}^{k_u} c_{ij}\bar{p}_{ij} + \gamma_{k_2}^k(q) \qquad (8.32)$$

where

$$\bar{p}_{ij} = p_i p_j \prod_{h=i+1}^{j+1}(1-p_h) \tag{8.33}$$

$$\gamma_{k_u}^k(g) = p_{k_u}(c_{i1}+c_{1i}) \sum_{l|\xi_{k_u}^l>g} p_{k_u}^l, (0<g\leq q) \tag{8.34}$$

$$\gamma_i^k(g) = (i-p_i)\gamma_{i+1}^k(g) + p_i\left[\sum_{l|\xi_{i+1}^l>g} p_i^l\left(\gamma_{i+1}^k\left(g-\xi_i^l\right)+c_{i1}+c_{1i}\right)\right.$$

$$\left. + \sum_{l|\xi_{i+1}^l<g} p_i^l \gamma_{i+1}^k\left(q-\xi_i^l\right)\right]$$

$$\Pr(\xi_i = g)\left[pi\left(\gamma_{i+1}^k(q) + \sum_{j=i+1}^{k_u} \bar{p}_{ij}(c_{i1}+c_{1i}-c_{ij})\right),\right. \tag{8.35}$$

$$\left.(i = k_2,\ldots,k_u-1; 0<g\leq q)\right]$$

$$T(x^v) = \sum_{k=1}^{|K|} T^k(x^v) \tag{8.36}$$

$$y \geq \widetilde{Q}(x^v)\left(\sum_{(i,j)\in E^v} x_{ij} - n + K + 2\right) \tag{8.37}$$

$$\widetilde{Q}(x^v) = T(x^v) - \tilde{c}x^v \tag{8.38}$$

Laporte et al. (2002) suggested a similar L-shaped stochastic programming algorithm for the SVRP-D, providing a valid lower bound on the value of $\widetilde{Q}(x^v)$; Laporte et al. (1992) adopted the previously mentioned algorithm to solve the SVRP-T, which considers a time window only at the depot and the expected penalty cost $\widetilde{Q}(x^v)$ refers to the amount of excess duration of a solution.

Recently, a branch and price algorithm has also been presented for the SVRP-D by Christiansen and Lysgaard (2007). Using the Dantzig–Wolfe decomposition, the SVRP-D has been decomposed into a set-partitioning master problem and a nonelementary shortest path problem with two-cycle elimination and has been solved as a subproblem. An upper bound ($S_{\max}$)

on the variance S_k of the demand distribution on a route k is determined using a 0-1 knapsack problem, which helps eliminate some of the non-elementary routes. Defining $F(\mu, \sigma^2, U)$ as the probability that the total actual demand along a partial path ($i_0 = 0, i_1,...,i_u$), with an expected demand distribution characterized by (μ, σ^2), does not exceed a positive integer U, the probability that the lth route failure will occur at customer i_u is given by $F(\mu - \mu_{i_u}, \sigma^2 - \sigma_{i_u}^2, lq) - F(\mu, \sigma^2, lq)$ (Dror et al. 1989). Using the recourse policy (b) (Section 8.4.2) and defining the expected total number of route failures at customer i_u by $Fail(\mu, \sigma^2, i_u)$ (Equation 8.39), Christiansen and Lysgaard (2007) calculated the expected route failure cost along the partial path as Equation (8.40):

$$FAIL(\mu,\sigma^2,i_u) = \sum_{l=1}^{\infty} F(\mu - \mu_{i_u}, \sigma^2 - \sigma_{i_u}^2, lq) - F(\mu, \sigma^2, lq)$$

$$EFC(\mu, \sigma^2, i_u) = 2c_{oi_u} FAIL(\mu, \sigma^2, i_u)$$

(8.39)

Defining an acyclic graph $G_S = (V_S, A_S)$ with vertices $v(\mu, \sigma^2, i)$, for $\mu = 1,..., q$, $\sigma^2 = 1,..., S_{max}$, and $i = 0,..., n$, Christiansen and Lysgaard (2007) constructed the arc set A_S and corresponding reduced costs as follows:

a. Add an arc from origin depot $v(0, 0, 0)$ to $v(\mu_i, \sigma^2_i, i)$ with the reduced cost $c_{0i} + EFC(\mu_i, \sigma^2_i, i) - \pi_i$, $\forall i \in C$, where π_i shows the dual variable (price) for customer i, obtained in the master problem.
b. Add an arc from $v(\mu, \sigma^2, j)$ to $v(\mu + \mu_i, \sigma^2 + \sigma^2_i, i)$ (provided that $\mu + \mu_i \leq q$, $\sigma^2 + \sigma^2_i \leq S_{max}$) with its reduced cost equal to $c_{ji} + EFC(\mu + \mu_i, \sigma^2 + \sigma^2_i, i) - \pi_i$, $\forall i, j \in V$, $\mu = 1,..., q$, $\sigma^2 = 1,..., S_{max}$.
c. Add an arc from $v(\mu, \sigma^2, j)$ to $v(\mu, \sigma^2, 0)$ with its reduced cost set to c_{j0}, $\forall j \in C$, $\mu = 1,..., q$, $\sigma^2 = 1,..., S_{max}$.

The shortest path subproblem was used to find the minimum reduced cost shortest path in the graph G_S, and added to the master problem LP. The augmented master problem LP is solved again giving new dual variables π_i and $\forall i \in C$, which are used to redefine the arc costs in set A_S. This process runs in cycles until the subproblem fails to provide a negative reduced cost column (path) at which the branching is resorted if the solution at hand is not an integer; otherwise, the integer solution gives the optimal solution of the SVRP-D. Christiansen and Lysgaard (2007) tested two branching strategies: The first one was based on the expected route demand μ and the second based on the flow variables x_{ij} and concluded that the former branching strategy performs better.

8.4.4.2 Heuristics Approaches

Earlier heuristics for the SVRP-D were based on some modification of the well known Clarke-White algorithm (Stewart and Golden 1983; Dror and Trudeau 1986). Stewart and Golden (1983) also proposed a generalized Lagrange multiplier (GLM) heuristics by Lagrangianly relaxing constraint Equation (8.20) in the chance constraint model and determining their best value by solving the relaxed problem as an m-TSP, whereas, in the recourse constraint model, the penalties were considered as known optimum value of the Lagrange multipliers. Later, some complex meta-heuristics were also presented, a few of which are described here.

Gendreau et al. (1996a) presented a tabu search-based heuristics, TABUSTOCH, which provides a heuristics solution of the same SVRP-CD model as was considered in their exact solution approach (Gendreau et al. 1995) (Section 8.4.4.1). The neighborhood was searched by removing some customers and inserting them immediately before or after their nearest neighbors. To account for the savings obtained by removing or inserting a customer in a route, some proxy estimations of the expected route cost (Equation 8.32) were suggested to accelerate the TABUSTOCH. To reduce the computation effort further, insertion cost of a customer into a route was approximated by a factor of the largest cost of removing a vertex from that route. Their computational results shows that the TABUSTOCH worked efficiently, producing optimal solutions for 89% of the instances tested in Gendreau et al. (1995) in less computation time.

Following the recourse policy (d) (Section 8.4.2), Yang et al. (2000) modeled the expected cost $(f_j(g))$ of a route from a node (say, j) onward using a dynamic programming recursion, where g represents the residual capacity after serving j, and b represents an additional penalty if a route failure occurs. If E_j represents the set of all possible load values that the vehicle can carry after serving j, then for each such $g \in E_j$, the expected cost $(f_j(g))$ was defined as

$$f_j(g) = \min \begin{bmatrix} c_{j,j+1} + \sum_{l:\xi^1 \leq g} f_{j+1}(g - \xi^1) p_{j+1,l} \\ + \sum_{l:\xi^1 > g} \{b + 2c_{j+1,0} + f_{j+1}(g + q - \xi^1) p_{j+1,l}\}, \\ c_{j,0} + c_{j,j+1} + \sum_{l=1}^{|l|} f_{j+1}(q - \xi^1) p_{j+1,l} \end{bmatrix}$$

They proved that in the absence of any limiting constraint, such as customers' time windows or a limit on the recourse cost, finding an optimal SVRP route is essentially a single vehicle routing problem. An insertion heuristics has been used to construct such a route, subject to an OR-opt local search scheme for improvement. Instead of using the forward dynamic programming for the whole route, to get the insertion cost of a sequence $(i_a,..., i_b)$ into a route between customers i and j, a proxy saving calculation is suggested, as shown in Equation (8.42). Here, $f_i^2(g)$ gives the expected route cost at i, calculated in backward recursion from j and through the sequence $(i_a,..., i_b)$, whereas $f_i^1(g)$ represents the corresponding cost before insertion:

$$\text{Proxy insertion cost} = \frac{\left[\sum_{g \in E_i}\left(f_i^2(g) - f_i^1(g)\right)\right]}{|E_i|}$$

Limiting the total recourse cost to T for a multivehicle SVRP-D with recourse policy (d), they suggested a route-first/cluster-next heuristics, which divides the single vehicle route (described before) into least cost feasible clusters or partitions. Their calculation results show that the route-first/cluster-next heuristics outperforms a cluster-first/route-next heuristics. Recently, Tan et al. (2007) developed a multiobjective evolutionary algorithm (MOEA) with three objectives of travel distance, number of vehicles, and driver remuneration for a SVRP-D that considers drivers' overtime if a route duration exceeds a limit (B). Their algorithm is an adaptation of their earlier work for deterministic VRPTW (Tan et al. 2003a, 2003b), except for the inclusion of a route simulation method (RSM) and local search exploitation. In the RSM, to evaluate the chromosomes (i.e., to calculate the corresponding expected routes' cost), a number of sets of customers' demands are generated at random to simulate the SVRP-D; these sets are refreshed after a fixed number of generations. The local search exploitation is based on two general properties for the SVRP-D given by Dror and Trudeau (1986): First, the probability of route failure is more at the end of route and, second, the route cost differs based on the direction of the route. Correspondingly, the first local search suggested was the shortest path search (SPS), which tries to resequence the customers in a route so that the farthest customers from the depot are served first in the route; the second one reverses the route direction.

8.5 THE STOCHASTIC VEHICLE ROUTING PROBLEM WITH TIME WINDOWS

As compared to the SVRP, considerably less attention has been given to the SVRPTW. One of the pioneering works in this field is due to Taniguchi, Yamada, and Tamagawa (1999), who proposed a probabilistic

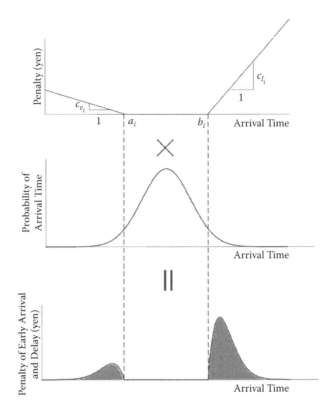

FIGURE 8.3 Early and late arrival penalty consideration in the P-VRPTW. (Source: adapted from Taniguchi, E. et al., 2001, *City Logistics: Network Modeling and Intelligent Transport Systems.* Oxford, England: Pergamon.

vehicle routing and scheduling problem with time windows (P-VRPTW). The P-VRPTW considers a travel time distribution for each arc of the instance, while minimizing a composite objective function consisting of the stochastic versions of Equations (8.1), (8.11), and (8.12). Figure 8.3 shows the consideration of the early and late arrival penalties in the P-VRPTW model.

Furthermore, instead of a single travel time distribution covering the uncertainty of travel time in a whole day, different travel time distributions were used for each arc in every hour. A macroscopic traffic simulator was used to accumulate historic data by simulating a certain number of days, providing the travel time distributions for every hour of a day. Taniguchi et al. (1999) also proposed a GA to solve the P-VRPTW model. Later, the P-VRPTW model was applied to much city logistics-related research, such as the environmental benefits of using stochastic travel times (Taniguchi et al. 2001), effects of e-commerce (Taniguchi and Kakimoto 2003), and travel time reliability of a road network (Yamada, Yoshimura, and Mori 2004). Recently, utilizing VICS data and the data from sixty-six days' operation of probe pickup/delivery trucks, Ando and Taniguchi (2007) have applied the P-VRPTW and its GA solution to evaluate an actual delivery system in Osaka, Japan.

Guo and Mak (2004) have considered soft time windows (Equation 8.11) in the objective function of an SVRP-CD model similar to the one given by Gendreau et al. (1995, 1996a). Their proposed GA to solve their SVRPTW-CD model is based on a heuristics edge-based crossover (HEC) and scramble sublist mutation (SSM). The HEC builds an offspring by selecting a customer i from a parent at random, and the next customer inserted in the offspring is either the successor of i in parent 1 or parent 2 based on their respective insertion costs. If both successors are infeasible, a random unrouted customer is inserted. In the SSM, a partial length of an individual is selected and it is replaced with a permutation of the customers in it. A comparison with another GA using partially mapped crossover and order-based mutation shows significantly better performance of their proposed GA. A comparison with an exact solution based on a branch and bound approach has also showed the efficiency of this proposed GA; however, no description of the exact branch and bound algorithm has been provided.

8.6 THE STOCHASTIC AND DYNAMIC VEHICLE ROUTING PROBLEM WITH TIME WINDOWS

The DVRPTW does not forecast any probable change in an uncertain data input and it always reacts to a changing input in the form of reoptimization when the change is actually realized. On the other extreme, the SVRPTW foresees all possible future changes considering the probability distribution of an uncertain data input but it (except the SVRP with recourse type c [Section 8.4.2]) fails to react to it actively. Rather, it relies on a predefined recourse policy. The stochastic and dynamic vehicle routing problem with time windows (SDVRPTW) lies at the juncture of the DVRPTW and the SVRPTW. It considers forecast future information based on the probability distribution or historical data about an uncertain input and includes its effect at the first stage of route planning; therefore, when the actual value of that input is realized, the routing plan has already some capacity to absorb this newly realized input value, resulting in better reoptimization.

As mentioned earlier, the DVRPTW can be used to model postdisaster scenarios; the SDVRPTW can be a better choice as the resulting routing plans already consider the forecast of near future events such as changes in the uncertain data (amount and location of the relief requirement, etc.) and thus are more robust. In the business logistics area, the SDVRPTW has many applications where dynamic decisions are required and where the future demand can also be predicted with some certainty, such as in heating oil distribution, dial-a-ride systems, etc. Relocation and grouping of taxis outside a railway station can be a good example of the SDVRPTW-oriented decision making.

Most of the research in the area of the SDVRPTW addresses the stochastic customers case (see Bent and Hentenryck 2004; Ichoua, Gendreau, and Potvin 2006) with some contributions in the area of stochastic demand (see Hvattum, Løkketangen, and Laporte 2006); no literature can be found that considers the stochasticity of the travel time in the SDVRPTW framework. The SDVRPTW can be modeled as two-stage (Hvattum et al. 2006) or multistage (Ichoua et al. 2006) stochastic programming recourse models where reoptimization is adopted as the recourse policy. However, in the algorithms, it has been represented using a rolling horizon model.

8.6.1 Waiting Strategies

In the literature, the wait-first waiting policy has usually been followed in the SDVRPTW (e.g., see Bent and Hentenryck 2004; Hvattum, Løkketangen, and Laporte 2007). Ichoua et al. (2006) proposed a different waiting strategy: If the probability of arrival of an immediate request near the current location of the vehicle is higher than a prespecified threshold, the vehicle waits for a certain amount of time (δt) that depends on the depot's and customers' time windows along the remaining part of the route. To avoid clustering of the waiting vehicles in a particular zone with high probability of arrival of immediate requests, an upper bound was assigned with each zone in the experimental setting.

8.6.2 Instances

Similarly to the cases of the DVRPTW and the SVRP, no benchmark instances are available for the SDVRPTW; researchers have worked with modified Solomon's benchmark instances (e.g., Bent and Hentenryck 2004) or with self-generated instances (see Hvattum et al. 2007; Ichoua et al. 2006). The key factors in generating/modifying Solomon's benchmark instances are the degree of dynamism, division of the geographical area, probability distributions of the stochastic customers, arrival rates, events, and division of the scheduling horizon. For example, Bent and Hentenryck (2004) have adopted the Solomon instances, dividing the scheduling horizon with the occurrence of events such as arrival of a new immediate request, vehicle departure, etc. Hvattum et al. (2006) have considered ten equal time intervals while generating customers' locations based on historical data of a real-life case. Ichoua et al. (2006) have divided the customers in two regions (simulating CBD and the peripheral areas) with different arrival rates in three different time intervals, whereas their rolling horizon settings are based on new customer arrival and service completion events.

8.6.3 Solution Approaches

For the most part, heuristics has been used to solve the SDVRPTW, relying on the generation of future scenarios—that is, using the similar techniques as the RSM in Tan et al. (2007) (Section 8.4.2.2). Hvattum et al. (2006) have commented that the solution produced using such sampling techniques could be nearly optimal when the sample size is large enough. Bent and Hentenryck (2004) presented a multiple scenario approach (MSA); many scenarios are created by sampling the probability distribution of the customers, and a greedy insertion heuristics is used to generate a routing plan for each scenario. These routing plans contain known as well as forecast stochastic customers; however, the stochastic customers are removed from these before saving them. An adaptive memory keeps these multiple routing plans; a consensus function is then used to return a *distinguished plan* based on the consistency with other plans, used for actual distribution. The adaptive memory and the distinguished plan are kept updated at the occurrence of dynamic events such as a new customer request, generation of a new plan, vehicle departures and timeouts, etc. Their computational results show that the MSA outperforms a similar multiple plans approach (MPA) that does not consider the stochastic customer information. Ichoua et al. (2006) modified the tabu search algorithm of Gendreau et al. (1999) (described in Section 8.3.3.2) in their vehicle-waiting heuristics by considering their waiting policy (Section 8.6.1). A dummy node is inserted in the route of a vehicle as the next destination if that vehicle is supposed to wait for duration of δt. It may be noted that the tabu search proposed by Gendreau et al. (1999) is similar to the MPA without the consensus function (Bent and Hentenryck 2004).

Hvattum et al. (2006) presented the dynamic stochastic hedging heuristic (DSHH) based on the sampling techniques mentioned earlier. The scheduling horizon is divided into equal time intervals and the DSHH finds a routing plan for each time interval at the start of each time interval including known customers (i.e., requests realized in previous time intervals) and forecast customers. The DSHH is a two-stage heuristic: The customers to be serviced in the next time interval are selected in the first stage and their order is decided in the second stage. Their algorithm also includes a concept similar to the consensus function of Bent and Hentenryck (2004), as the customers need to be served in a particular time interval are decided based on their consistent appearance as the first customers to be served in solutions of the sample scenarios. Hvattum et al. (2006) showed that the solution quality (in terms of minutes of travel time) increases with the increase in the number of sample scenarios solved per iteration of the heuristics with a linear increase in the computational effort. Their calculation results also show that the DSHH performs better than a pure DVRPTW heuristic (named myopic dynamic heuristics [MDH] in Hvattum et al. 2006) as far as the travel time is concerned, but the average number of required vehicles in the DSHH is more compared to the MDH. They conclude that this may be due to the fact that stochastic information leads to the commitment of an extra vehicle to cover the probable (but not yet realized) demands when the vehicle capacity is slightly higher than the current actual demands; on the other hand, the MDH does not consider any future information and produces plans with a minimum number of vehicles required to serve the already realized demands.

To overcome this drawback of the DSHH, Hvattum et al. (2007) presented a branch-and-regret heuristic (BRH) for the SDVRPTW with stochastic customers with/without stochastic demands. While the basic scheme of the BRH is the same as the DSHH, it considers both options (as two branches) of whether to service a customer in the next time interval (as in the DSHH) or not and choosing the best branch; also, its consensus function evaluates the two branches of serving or not, a customer first-up on a particular vehicle route. While the BRH was able to fix the problem of using more vehicles compared to the MDH, it resulted in slightly worse solutions in terms of travel time compared with the DSHH. For the case of SDVRPTW with stochastic customers and stochastic demands, the MDH using the expected value of the stochastic demand of known customers performed better than the BRH, with stochastic information on demand in terms of travel time, but with a slightly higher number of vehicles required.

REFERENCES

Alvarenga, G. B., Silva, R. M., and Mateus, G. R. (2005). A hybrid approach for the dynamic vehicle routing problem with time windows. *Proceedings of the Fifth International Conference on Hybrid Intelligent Systems (HIS'05) IEEE.*

Ando, N., and Taniguchi, E. (2006). Travel time reliability in vehicle routing and scheduling with time windows. *Networks and Spatial Economics* 6:293–311.

Bent, R. W., and Hentenryck, P. V. (2004). Scenario-based planning for partially dynamic vehicle routing with stochastic customers. *Operations Research* 52:977–987.

Bertsimas, D. J., Jaillet, P., and Odoni, A. R. (1990). A priori optimization. *Operations Research* 38:1019–1033.

Branchini, R. M., Armentano, V. A., and Løkketangen, A. (2009). Adaptive granular local search heuristics for a dynamic vehicle routing problem. *Computers and Operations Research* 36:2955–2968.

Chen, Z., and Xu, H. (2006). Dynamic column generation for dynamic vehicle routing with time windows. *Transportation Science* 40:74–88.

Christiansen, C. H., and Lysgaard, J. (2007). A branch-and-price algorithm for the capacitated vehicle routing problem with stochastic demands. *Operational Research Letters* 35:773–781.

Christofides, N., and Eilon, S. (1969). An algorithm for the vehicle dispatching problem. *Operational Research Quarterly* 20:309–318.

Dantzig, G. B., and Ramser, J. H. (1959). The truck dispatching problem. *Management Science* 6:80–91.

Dror, M., Laporte, G., and Trudeau, P. (1989). Vehicle routing with stochastic demands: Properties and solution framework. *Transportation Science* 28:166–176.

Dror, M., and Trudeau, P. (1986). Stochastic vehicle routing with modified savings algorithm. *European Journal of Operational Research* 23:228–235.

Fleischmann, B., Gietz, M., and Gnutzmann, S. (2004). Time-varying travel times in vehicle routing. *Transportation Science* 38:160–173.

Fleischmann, B., Gnutzmann, S., and Sandvoss, E. (2004). Dynamic vehicle routing based on online traffic information. *Transportation Science* 38:420–433.

Gendreau, M., Guertin, F., Potvin, J., and Taillard, E. (1999). Parallel tabu search for real-time vehicle routing and dispatching. *Transportation Science* 33:381–390.

Gendreau, M., Laporte, G., and Seguin, R. (1995). An exact algorithm for the vehicle routing problem with stochastic demands and customers. *Transportation Science* 29:143–155.

——— (1996a). A tabu search heuristics for the vehicle routing problem with stochastic demands and customers. *Operations Research* 44:469–477.

——— (1996b). Stochastic vehicle routing. *European Journal of Operational Research* 88:3–12.

Guo, Z. G., and Mak, K. L. (2004). A heuristic algorithm for the stochastic vehicle routing problem with soft time windows. *Evolutionary Computation, 2004, CEC2004. Congress on Publication*, 2:1449–1456.

Haghani, A., and Oh, S. C. (1996). Formulation and solution of a multicommodity, multimodal network flow model for disaster relief operations. *Transportation Research Part A* 30:231–250.

Hvattum, L. M., Løkketangen, A., and Laporte, G. (2006). Solving a dynamic and stochastic vehicle routing problem with a sample scenario hedging heuristic. *Transportation Science* 40:421–438.

——— (2007). A branch-and-regret heuristic for stochastic and dynamic vehicle routing problems. *Networks* 49:330–340.

Ichoua, S., Gendreau, M., and Potvin, J. (2000). Diversion issues in real-time vehicle dispatching. *Transportation Science* 34:426–438.

——— (2003). Vehicle dispatching with time-dependent travel times. *European Journal of Operational Research* 144:379–396.

——— (2006). Exploiting knowledge about future demands for real-time vehicle dispatching. *Transportation Science* 40:211–225.

Laporte, G., Louveaux, F., and Mercure, H. (1992). The vehicle routing problem with stochastic travel times. *Transportation Science* 26:161–170.

——— (2002). An integer L-shaped algorithm for the capacitated vehicle routing problem with stochastic demands. *Operations Research* 50:415–423.

Larsen, A. (2001). The dynamic vehicle routing problem. PhD thesis, LYNGBY, IMM-PHD-2000-73, IMM, Technical University of Denmark.

Larsen, A., Madsen, O., and Solomon, M. (2002). Partially dynamic vehicle routing—Models and algorithms. *Journal of Operations Research Society* 53:637–646.

Mitrovic-Minic, S., and Laporte, G. (2004). Waiting strategies for the dynamic pickup and delivery problem with time windows. *Transportation Research Part B* 38:635–655.

Ozdamar, L., Ekinci, E., and Kucukyazici, B. (2004). Emergency logistics planning in natural disaster. *Annual of Operations Research* 129:217–245.

Powell, W., Jaillet, P., and Odoni, A. (1995). Stochastic and dynamic networks and routing. In *Network routing*, ed. M. Ball, T. Magnanti, C. Monma, and G. Nemhauser, 141–295. Amsterdam: Elsevier Science.

Psaraftis, H. N. (1995). Dynamic vehicle routing: Status and prospects. *Annual of Operations Research* 61:143–164.

Qureshi, A. G., Taniguchi, E., and Yamada, T. (2009). An exact solution approach for vehicle routing and scheduling problems with soft time windows. *Transportation Research Part E* 45:960–977.

——— (2012a). Exact solution for vehicle routing problem with soft time windows and dynamic travel time. *Asian Transport Studies* 2:48–63.

——— (2012b). A microsimulation based analysis of exact solution of dynamic vehicle routing with soft time windows. *Procedia—Social and Behavioral Sciences* 39:205–216. (http://dx.doi.org/10.1016/j.sbspro.2012.03.102)

Sheu, J. B. (2007). An emergency logistics distribution approach for quick response to urgent relief demand in disasters. *Transportation Research Part E* 43:687–709.

Solomon, M. M. (1987). Algorithms for the vehicle routing and scheduling problem with time windows constraints. *Operations Research* 35:254–265.

Stewart, W. R., Jr., and Golden, B. L. (1983). Stochastic vehicle routing: A comprehensive approach. *European Journal of Operational Research* 14:371–385.

Tan, K. C., Cheong, C. Y., and Goh, C. K. (2007). Solving multiobjective vehicle routing problem with stochastic demand via evolutionary computation. *European Journal of Operational Research* 177:813–839.

Tan, K. C., Lee, T. H., Chew, Y. H., and Lee, L. H. (2003a). A hybrid multiobjective evolutionary algorithm for solving truck and trailer vehicle routing problems. *Proceedings of the 2003 Congress on Evolutionary Computation*, Canberra, Australia, December 8–12, 3:2134–2141.

——— (2003b). A multiobjective evolutionary algorithm for solving vehicle routing problem with time windows. *Proceedings of the IEEE International Conference on Systems, Man and Cybernetics*, Washington, DC, October 5–8, 1:361–366.

Taniguchi, E., and Kakimoto, Y. (2003). Effects of e-commerce on urban distribution and the environment. *Journal of the Eastern Asia Society for Transportation Studies* 5:2355–2366.

Taniguchi, E., and Shimamoto, H. (2004). Intelligent transportation system based dynamic vehicle routing and scheduling with variable travel time. *Transportation Research Part C* 12:235–250.

Taniguchi, E., Thompson, R. G., Yamada, T., and van Duin, R. (2001). *City logistics: Network modeling and intelligent transport systems.* Oxford, England: Pergamon.

Taniguchi, E., Yamada, T., and Tamagawa, D. (1999). Probabilistic vehicle routing and scheduling on variable travel times with dynamic traffic simulation. In *City logistics I,* ed. E. Taniguchi and R. G. Thompson, 85–99. Institute for City Logistics, Japan.

Yamada, T., Yoshimura, Y., and Mori, K. (2004). Road network reliability analysis using vehicle routing and scheduling procedures. In *Logistics systems for sustainable cities,* ed. E. Taniguchi and R. G. Thompson, 111–122. Bingley, UK: Elsevier.

Yang, W. H., Mathur, K., and Ballou, R. H. (2000). Stochastic vehicle routing problem with restocking. *Transportation Science* 34:99–112.

CHAPTER 9

Urban Transport and Logistics in Cases of Natural Disasters

Sideney A. Schreiner, Jr.

CONTENTS

9.1 Introduction	225
9.1.1 Disaster Classification	226
9.1.2 Stakeholders Affected by Disasters	227
9.2 Disaster Impact Analyses	227
9.2.1 Disaster Impact on the Infrastructure	228
9.2.2 Disaster Impacts on the Transportation Operation	229
9.2.3 Reconstruction Impacts	231
9.2.4 Disaster Risk Measurement	232
9.2.4.1 Quantitative Risk Measurement	233
9.2.4.2 Qualitative Risk Estimation	234
9.3 Business Continuity for Transportation	236
9.3.1 Facility Location	241
9.4 Conclusion	241
References	241

9.1 INTRODUCTION

Defining disasters is not a simple task. Leaning (2008) proposes a broad concept that disasters are complex phenomena that create extensive consequences for human populations. According to the same author, there is, however, a general agreement on the characteristics that make an event a disaster: A disaster is an event that imposes severe and intense stress on a community that cannot be dealt with through deployment of the ordinary resources of that community. Disasters are events that require outside help. Time is also considered in some accepted definitions of disaster: Disasters are events that contain anticipatory warning signs (although often not observed or looked for) but impose a relatively abrupt impact. The consequences of that impact may persist for months or years. Moreover, most definitions focus on impacts on and consequences for human populations. Events such as earthquakes and floods that do not cause direct or indirect social, economic, or psychological disruptions are usually not registered as disasters and not counted in disaster databases.

Natural disasters have resulted in insurmountable economic and human losses since the beginning of human history. Although individual and group preparations have been tried in order to mitigate the worst effects, the intensity of such disasters and their respective impacts are hardly predictable.

9.1.1 Disaster Classification

There is no official classification for disasters. The Center for Research on the Epidemiology of Disasters (CRED) keeps a publically available database on the occurrences and trends of disasters. Data on disasters over the several last decades have indicated that the incidence of disaster, despite divergence about its definition, is increasing (CRED 2007).

According to the disaster definition considered by CRED, disasters are categorized as natural disasters and technological disasters. In addition to the mentioned categorization, the most commonly used classification considers also its duration (short term or long term), its frequency of occurrence (frequent, eventual, or rare), and its intensity (high, medium, or low).

Except for the preparedness planning from the government side, the origin of the disaster is of minor importance. The government can directly address technological disasters by diverse political actions (governance, security, foreign policy, etc.). The private sector commonly addresses disaster preparedness by estimating and managing the imposed risk. In that sense, the characteristics that are commonly considered are the frequency, intensity, and impact (usually the loss in financial terms).

Specifically focusing on transportation, the interruption of a traffic network resulting from a disaster, or any other cause seriously influences both passengers and logistical distribution networks. Since logistical distribution is linked with local, regional, and even international economic activities, the damage on the freight system is as serious as that on passenger transportation. In that sense, the classification of the impacts of the disasters on each infrastructure—or perhaps the classification of each infrastructure in terms of its local, regional, or international socioeconomic importance—should be prioritized rather than the disaster classification itself.

TABLE 9.1 Stakeholders and Impacts of Disasters

Stakeholder	Immediate Impact	Medium and Long-Term Impacts
General population	Deaths	
	Injuries	Psychological disruption
	Property loss	Economic loss
Government	Social disorder	Reconstruction costs
	Economic loss	Economic reactivation costs
Private sector	Workforce disruption	Supply chain and operation disruption
	Property loss	

9.1.2 Stakeholders Affected by Disasters

Technological disasters tend to be focalized in small areas, while natural disasters usually achieve catastrophic proportions, with impacts in practically all sectors of society. Table 9.1 shows an overview of the impacts of natural disasters on different stakeholders.

Stakeholders perform different roles during response to a natural disaster. Regardless of these roles, disaster response is closely connected to transportation and logistics planning. During the immediate disaster response, the key activities that play a role in ensuring the maximum number of survivors are rescue, evacuation, and relief distribution. Such activities, referred to as humanitarian logistics, are subject to ongoing research and will be addressed in another opportunity. During the postdisaster reconstruction period, the efficient social and economic recovery in the affected area is ensured by rehabilitation and reconstruction, which are also closely related to transportation and logistics planning.

9.2 DISASTER IMPACT ANALYSES

The impact of natural disasters on the transportation system has two components: the direct impact on the infrastructure and the indirect impact on the operation, which is a result of the sudden reduction in network capacity and immediate changes in the demand, such as introduction of reconstruction vehicles in the network, inflow of volunteers, and evacuating the population.

The direct impact of disasters can be addressed by considering the vulnerability of the population and infrastructure. The indirect impact on transportation operation is highly dependent on the direct impact, mainly with regard to the compromised infrastructure. The relocation of population through evacuation causes alterations of the travel demand. These alterations can be temporary, such as in the case of floods, or permanent in cases of severe earthquakes, for example. These two components are presented in detail in the following sections.

9.2.1 Disaster Impact on the Infrastructure

The impact on the infrastructure is a result of the structural vulnerability of the transport system's physical elements, such as pavement, bridges, tunnels, etc. The estimation of such vulnerability usually considers the intensity required in order to damage or to collapse an infrastructure of a certain material and age. Certain types of disasters, such as floods, may put the infrastructure out of service while not impacting its physical elements. The monetary consequences related to this type of impact are expressed in clearing, repair, or reconstruction costs. This type of impact is considered a direct impact of the natural disaster that is felt mainly by the government and the entities responsible for network maintenance and reconstruction.

Resilience (conceptually, the opposite of vulnerability) of the transport system can be increased partially by ensuring appropriate construction standards that cover most frequent disasters in the area where the network is located. However, due to the unpredictable nature of the disasters, neither collection of statistical data nor disaster-proof construction can be taken as rule in most situations. Thus, assuming that the disasters will eventually compromise some infrastructure elements, it is important to understand the size of the impact on the transportation system when such infrastructure elements fail.

The concept of critical infrastructure is one way to evaluate the importance of each infrastructure element. Although the strict definition of critical infrastructures is still being discussed, the main idea serves for the objective at hand.

According to the International Risk Government Council (IRGC 2006), critical infrastructures are "a network of independent, large-scale, man-made systems (set of hard and soft structures) that function collaboratively and synergistically to produce a continuous flow of essential goods and services." Eusgeld et al. (2009) add that such infrastructures are essential for economic development and social well-being and that they are subject to multiple, potentially asymmetrical threats (technical, intentional or unintentional human, physical, natural, cyber, contextual) and may pose risks themselves. The concept has been gradually broadened from its original meaning, which is "those structures whose prolonged disruption could cause significant military and economic dislocation" (Mote, Copeland, and Fischer 2003). The German Federal

Agency for Security in Information Technic (Bundesamt für Sicherheit in der Informationstechnik [BSI]), for example, now defines critical infrastructures as "organizations or facilities of key importance to public interest whose failure or impairment could result in detrimental supply shortages, substantial disturbance to public order or similar dramatic impact" (BSI 2004).

Both of these definitions are broad and are at the same time of a "counterfactual" nature. They are broad because they cover sectors such as transportation, energy, hazardous materials, telecommunications, finance/insurance, and, of course, services such as waste management, water and food, civil protection, and health. They are also broad because it is typically assumed that "if individual such infrastructures are affected by deliberate disruption (information warfare, cyber terror, etc.) or failures of information technology, this could have the effect of triggering a chain reaction of disruption in other areas as well" (BSI 2004). The notion of critical infrastructure is counterfactual because we do not know in advance, due to systems evolution, complexity, shifting intentions, and actions of relevant actors, where failure will occur and what impact this will have. Critical infrastructure then risks becoming everything that we think is important to the functioning of society and yet nothing really specific enough to focus vulnerability reduction measures on. The fact is that in these vast sociotechnological systems, failure is indeed a normal condition (Perrow 1984).

In the U.S. National Strategy for the Physical Protection of Critical Infrastructures and Key Assets (2003), a systemic "effect-vulnerability" notion of critical infrastructure is proposed: namely, that three types of effects may confer vulnerability on a system:

- Direct infrastructure effects: cascading disruption or arrest of the functions of critical infrastructures or key assets through direct attacks on a critical node, system, or function
- Indirect infrastructure effects: cascading disruption and financial consequences for government, society, and economy through public- and private-sector reactions to an attack
- Exploitation of infrastructure: exploitation of elements of a particular infrastructure to disrupt or destroy another target

9.2.2 Disaster Impacts on the Transportation Operation

The impact of a natural disaster on the operation of a transport system is the consequent increase in travel time and costs due to the reduction of capacity for the period between the disaster occurrence and

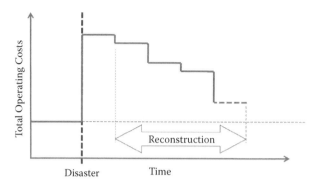

FIGURE 9.1 Disaster impact on the operational costs of transportation.

the completion of reconstruction work (Figure 9.1). Other costs, such as the loss of equipment and changes in the demand behavior following a large-scale natural disaster, also impact the transportation operation. However, as such changes are closely related to the habits of the affected population, this will not be analyzed in this study.

The impact on travel time and costs is considered to be an indirect impact of the natural disasters. This is, however, the type of impact that can be diminished by implementation of response plans for both passenger and freight transport. The efficiency of the response plans relies a great deal on the adequate availability of information after the disaster, especially regarding the condition of the network. The duration of such impacts on the operation is dictated by the reconstruction process, which will define which network elements will be reconstructed and when the reconstruction will take place. The reconstruction operation planning during the postdisaster period should take into consideration that the reconstruction process can potentially contribute to diminishing the disaster's impact on the operation.

Several other indirect impacts have been observed. Depending on the point of view, each stakeholder should also consider different impacts:

- Financial loss: for example, the loss of orders for a period of time, additional costs to recover service, loss of market share, etc.
- Reputational damage: for example, loss of goodwill or credibility, political or corporate embarrassment, compromised health and safety, etc.
- Legal action: for example, contractual breaches, personal details being made public and infringing data protection legislation,etc.

9.2.3 Reconstruction Impacts

After a large-scale natural disaster in a densely populated area, humanitarian logistics activities will follow. These activities will include rescue, relief distribution, and emergency reconstruction. During this period, normal commercial activities are usually disrupted or set on hold. Once this initial stage has passed, reconstruction of destroyed buildings and utilities networks starts. Although there is no clear moment when humanitarian logistics activities give space to reconstruction, since reconstruction may start in some regions while rescue is still underway at other locations, this study considers that the reconstruction stage has a specific initial moment. The problems of reconstruction scheduling, distribution of reconstruction material, and traffic regulation after a natural disaster are addressed by Finn, Ventura, and Schuster (1995), Yamada et al. (1986), Wu (1997), Chen and Tzeng (1999), Chen (1999), and Chang (2001), among others. In practice, decision makers must assign and dispatch manpower, machines, vehicles, and materials to rebuild the damaged infrastructures according to clear objectives, such as maximizing the number of lives saved during rescue, maximizing transport accessibility, and minimizing the risk during the reconstruction. Although the literature focuses on accessibility of general traffic, highly industrialized areas should consider freight accessibility as an important measure to ensure the continuity of a local economy.

The reconstruction of damaged networks affects both the infrastructure and the operation. Feng and Wen (2005) point out the variability in the time necessary for the reconstruction of damaged transport networks by mentioning four examples (see Table 9.2).

The adequate reconstruction planning must consider two stages of reconstruction: emergency rehabilitation and reconstruction. Each stage is defined by different objectives and available resources. During the emergency rehabilitation, maximization of rescue efficiency and protection of compromised critical infrastructure are the objectives, restricted by extremely limited resource availability. During the reconstruction

TABLE 9.2 Reconstruction Time of Damaged Transportation Networks

Region	Reconstruction Time
Loma Pietra, San Francisco, US	3 Months
Northridge, Los Angeles, US	3 Months (interstate); 10 months (local)
Hyogoken-Nanbu, Kobe, Japan	12 Months
Chi-Chi, Taiwan	24 Months

Source: Feng, C. M., and Wen, C. C. (2005). *Journal of the Eastern Asia Society for Transportation Studies* 6:4253–4268.

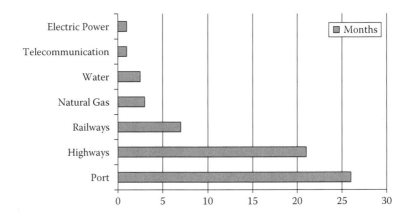

FIGURE 9.2 Reconstruction time for lifeline networks in Kobe. (Source: Chang, S. E., 2000, Transportation Performance, Disaster Vulnerability, and Long-Term Effects of Earthquakes. Working draft.)

period, quality of the rebuilt infrastructure and minimization of reconstruction costs are the usual objectives, while resource availability is less restrictive (Feng and Wen 2005).

The main impact of the reconstruction on the operation is the modification of travel times and travel costs during the reconstruction period as a result of the gradual restoration of capacity. This impact, although mostly irrelevant to the public, is considerable to freight transport, which will operate in nonoptimal scenarios. Figure 9.2 shows the relative disparity in the time necessary for the reconstruction of different lifeline infrastructures in Kobe after the Hyogoken-Nanbu earthquake.

9.2.4 Disaster Risk Measurement

The term "risk" has been utilized in different contexts. It usually relates to safety and security and is also applied to business investment as business risk or project risk, among others. Risk usually refers to something that is uncertain and that uncertainty normally incurs a loss or adverse effect. The concept of *risk management* emphasizes the systematic identification, analysis, and assessment of all hazards inherent to an activity so that effective measures can be established to control the risks.

An important aspect for the determination of risk is the assessment of the worst-case scenario given a certain disaster intensity. This will improve the chances for a more successful response due to better planning prior to the actual disaster. The existence of efficient tools and data for the predisaster plan is a key factor for the improvement of evacuation, rescue, and recovery after the disaster. From the point of view of people and freight transportation, the proper evaluation of the risk is the key to successful response plans.

9.2.4.1 Quantitative Risk Measurement

For quantitative measurement, risk is normally considered to be defined by the relation between the probability of the disaster occurrence and its consequences, or between the probability of occurrence and the vulnerability of the affected system. In order to propose an evaluation method to a specific system—in this case, a transportation-based business—risk is considered for specific scenarios, where the disaster intensity and its damage are known and presented in a mathematical formulation by Kaplan and Garrick (1981). According to their formulation, risk is suggested in terms of a set of triplets:

$$R = \{<s_i, p_i, x_i>\} \quad i = 1, 2, \ldots N$$

where
i is the studied scenario
N is the number of analyzed scenarios
s_i is a scenario identification
p_i is the probability of that scenario
x_i is the evaluation measure of damage of that scenario

This definition implies that scenarios should be created in an attempt to relate the cause of the damage, usually known as hazard; the probability of occurrence; and the expected damage. The first observation to be made is that different hazards can cause similar damage, with different occurrence probabilities. This allows one to focus on the effects of different disasters, rather than causes.

The evaluation of the risk is then formulated as in Equation 9.1:

$$R_i = P(i) \cdot X_i \qquad (9.1)$$

where
R_i is the quantitative representation of the risk in scenario i
$P(i)$ is the probability of occurrence of scenario i, which relates a hazard, an intensity, and a frequency
X_i is the expected damage to the system in study when scenario i happens

In the analysis, calculated or estimated values of probabilities and consequences have to be obtained. Based on these values, an evaluation is then made as to whether the risk is acceptable or safety measures are necessary. The use of this equation presumes knowledge of incident probability that is far more extensive and precise than seems to exist in reality. However, subjective judgments have to be used in many situations to determine the likelihood of the occurrence of an event as well as its severity. Therefore, quantitative risk assessment is generally used when historical data are available.

Equation 9.1 also presumes the quantification of the consequences of the disasters. The assessment of such consequences is based on the identification of the affected stakeholders and the impacts suffered. Specifically regarding the impacts on the transportation systems, the reduction of network capacity and traffic demand alterations occur in consequences represented by increase of travel times. Moreover, depending on the stakeholder, damage to vehicles and owned infrastructure is considered in the reduction of operating capacity.

9.2.4.2 Qualitative Risk Estimation

There are also systematic approaches in estimating risk qualitatively by using appropriate decision-making tools, often when historical data are not available. One commonly used approach is the development of a "risk assessment decision matrix," which is adopted from the U.S. Military Standard System Safety Program Requirements, known as MIL-Std-882-B. This type of matrix can be employed to measure and categorize risks on an informed judgment basis as to both probability and consequence and as to relative importance. An adaptation of the matrix is illustrated in Figure 9.3.

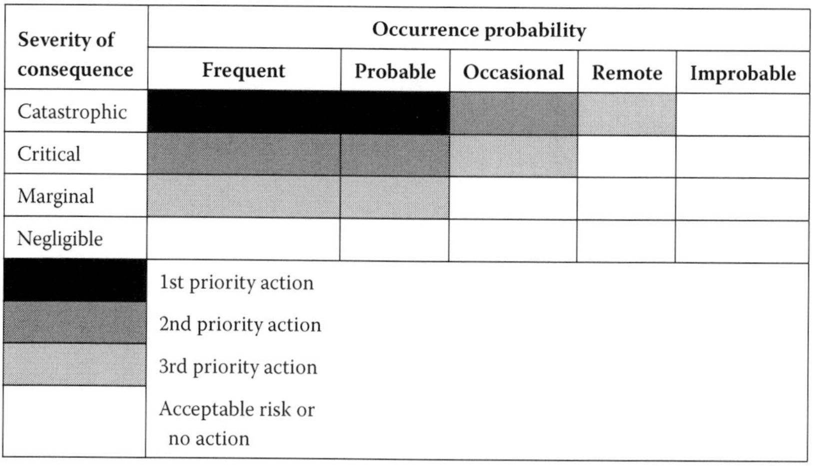

Severity of consequence	Occurrence probability				
	Frequent	Probable	Occasional	Remote	Improbable
Catastrophic					
Critical					
Marginal					
Negligible					

■	1st priority action
▓	2nd priority action
░	3rd priority action
	Acceptable risk or no action

FIGURE 9.3 Example of risk matrix.

The matrix presented in the figure shows that risks can be roughly classified into different categories depending on the parameters on the two axes. The different categories of risks require different ranks of control action. Priority on control actions can be determined by the adoption of this kind of decision-making tool. According to Yeh (1999), the most common risk management strategies can be divided into three categories: retention, mitigation, and transfer, which are described in detail next.

Risk retention refers to risk that is borne by the sufferer alone due to the inability to transfer or impossibility of transferring that risk to others. In the risk retention principle, the focus is on how best to absorb the risk by oneself, to reduce the impact of the disaster effectively, before or after its occurrence (Yeh 1999). In practice, several types of industries often reserve some capital, also designated as reserve funds, to permit the smooth management of emergency situations.

Risk transfer refers to various methods used to shift or transfer the risk to others. For example, efforts may be made to try to reduce risk loss by purchasing an insurance policy; by the mechanism of industrial coinsurance, in which the interested parties enter into a strategic alliance to share the risk; by integrating the local tourism industry to obtain capital with the sale of catastrophe bonds; and so on (Hsu et al. 2006).

Currently, the most commonly used disaster risk transfer tools (financial tools) in domestic and international markets are catastrophe insurance and catastrophe bonds. For example, the Tokyo Disney Resort issued a catastrophe bond to help it cope with diverse earthquake risks such as occurred in the Tokyo region in 1999. The program is designed to transfer the earthquake risk by enhancing underwriting capability while obtaining credit through capital markets (Chang 2000; Tsai and Chen 2009). Taking into consideration the characteristics of the insured industry, other types of risk transfer tools, such as weather insurance, and industrial coinsurance mechanisms are also very suitable. Another type of transfer tool is parametric insurance, which specifies an event with specific intensity. For example, it is possible to obtain insurance for a company building referring to a magnitude 7 or higher earthquake with its epicenter at a location within a 50 km radius from the building. In case the event happens within the agreed conditions, the company receives the payment regardless of damage to the building.

A major part of the disaster recovery planning process is the assessment of the potential risks to the organization that could result in the disasters or emergency situations themselves. It is necessary to consider all the possible incident types, as well as the impact that each may have on the organization's ability to continue to deliver its normal business services.

9.3 BUSINESS CONTINUITY FOR TRANSPORTATION

Business continuity is the ability to maintain production during and after disasters. It is based on the creation and validation of a practiced logistical plan, the business continuity plan (BCP), for how an organization will recover and restore partially or completely interrupted critical (urgent) functions within a predetermined time after a disaster or extended disruption. Disasters include local incidents like building fires, regional incidents like earthquakes, or national incidents like pandemic illnesses.

The economic disruption caused by damage to lifeline systems such as electric power and transportation has been modeled in different methodologies (Rose et al. 1997; Cho et al. 2000). However, these models assume that long-term economic impacts do not occur. Chang (2000) presented empirical evidence that long-term economic impact occurred in Kobe, after the Hyogoken-Nanbu earthquake, due to damage to transport infrastructure. In Kobe, the earthquake caused severe damage to the regional transportation system, including highways, regional railways, high-speed inter-regional rail, and seaport facilities. In that disaster, transport systems required much longer time frames for repair than other urban lifeline systems. The interruption of those systems alongside the direct impact to organizations' infrastructure and personnel caused partial and total discontinuity. The port of Kobe, as presented in Figure 9.4, had an immediate drop in its market share of imports in

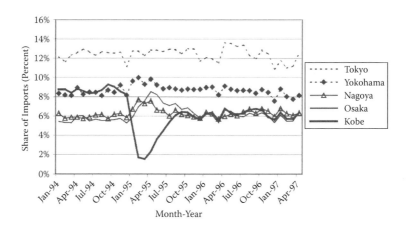

FIGURE 9.4 Effect of Hyogoken-Nunbu earthquake on the market share of imports in Japan. (Source: Chang, S. E., 2000, Transportation Performance, Disaster Vulnerability, and Long-Term Effects of Earthquakes. Working draft.)

Japan, which never recovered to the same level even after reconstruction was completed.

Business continuity management (BCM) is a management tool that can be employed to provide greater confidence that the outputs of processes and services can be delivered in the face of risks. It is concerned with identifying and managing the risks that threaten to disrupt essential processes and associated services, mitigating the effects of these risks, and ensuring that recovery of a process or service is achievable without significant disruption to the enterprise.

Gibb and Buchanan (2006) present a framework for the development of a business continuity plan based on different approaches proposed by several authors. This framework contains nine phases covering specific activities, inputs, and outputs, as shown in Table 9.3.

In a summarized perspective, the answers to the four following questions should be addressed by the implementation of BCM strategies:

- What is the worst thing that could happen to our business?
- Where would we be operating tomorrow if a disaster occurred?
- How quickly could our business reach the point of no return?
- How quickly can we return to business as usual?

Such questions represent the essence of disaster-related continuity planning, identifying the threats to the business, identifying backup strategies, measuring vulnerability, and measuring the efficiency of response plans.

Transportation can be disrupted by a large-scale natural disaster in two ways. The first is the disruption of passenger transportation, which in urban areas generally means that operation of railway systems (metro or urban rail) and bus systems (conventional buses or bus rapid

TABLE 9.3 Phases of a Business Continuity Plan

Phase	Description
1	Program initiation
2	Project initiation
3	Risk analysis
4	Selecting risk mitigation strategies
5	Monitoring and control
6	Implementation
7	Testing
8	Education and training
9	Review

transits) is affected in a way that people are unable to reach their usual destinations such as school or workplace.

The second way in which transportation is affected is the disruption of freight transportation. Naturally, the impact of such disruption is the reduction of goods supply for the population as a final client. However, the whole production chain is affected, especially manufacturing companies depending on transportation of goods between suppliers and production facilities, and all other distribution industries, such as fuel, food, etc.

In reality, the transportation disruption can affect different links of the supply chain: transportation issues on the input of, during, and on the output of the production process. The inability to receive raw materials, generally speaking; the inability to transfer components or goods from one factory to the next in a decentralized assembly system; and the inability to distribute products to clients reflect the potential for transportation discontinuity.

In order to ensure the continuity of transportation activities after a large-scale natural disaster, it is possible to apply the definition of business continuity management to transportation. In this sense, phases 3, 4, 5, and 6 of the framework presented by Gibb and Buchanan (2006) can be modified to ensure the continuity of a transportation-based business.

In phase 3, it is necessary to consider that, even though the risk analysis should cover the probability of a disaster in a predefined intensity and the impact on the transportation operation, most companies will not be able to evaluate the probability of the disasters in a straightforward way. An alternative to the evaluation of the risk posed by disruption of a critical element of the transportation infrastructure is the impact evaluation of total (or partial) disruption of the company-specific critical infrastructure.

Therefore, it is necessary to introduce the concept of a company-specific critical infrastructure, which is the group of all infrastructure elements that will significantly reduce accessibility between suppliers, depots, assembly or processing units, and clients for that specific company. Quantitatively, such reduction in accessibility should be considered in two dimensions: absolute impact in accessibility and duration of accessibility reduction. That means that the total impact on operating costs should consider an analyses horizon that covers the period between the occurrence of the disaster and the completion of reconstruction work.

It is also necessary to define general transportation critical infrastructure, which is identified by its structural position in the network (trunk links), by its high capacity, or by the long time required to reconstruction (tunnels and bridges). All of these characteristics are reflected as significant increases in travel time and travel costs in case of partial or total disruption.

The evaluation of the accessibility reduction can be performed using a modified version of the methodology presented by Chang and Nojima (2001). In this model, the accessibility of a location k at a time t after the disaster ranges from zero (complete accessibility loss) to one (no loss) and is given by Equations 9.2 and 9.3. The general accessibility of the region where the facility is located is given by Equation 9.2, while the accessibility between specific locations of two facilities can be evaluated by Equation 9.3.

$$A_k(t) = \frac{1}{n_k} \sum_{i \in N_i} A_i(t) \qquad (9.2)$$

where
$A_k(t)$ is the accessibility ratio of region k at time t
i indicates a node on the transport network
n_k is the number of nodes in k

In order to incorporate partially disrupted roads in the calculation of Equation 9.3, the links are required to have an individual postdisaster equivalent distance that incorporates the increase in travel times:

$$A_i(t) = \frac{\sum_{j \neq i} w_{ij} d(t)}{\sum_{j \neq i} w_{ij} d(t_0)_{ij}} \qquad (9.3)$$

where
N_k is the set of nodes in k
d_{ij} is the minimum distance on the damaged network from node i to node j
$d(t_0)$ is the minimum distance on the predisaster network (t_0) from node i to node j
w_{ij} is the weight for node j for shipments from or to node i

Although the original model uses a multiplier to the original length of the link to compute the increase in travel time due to the partial interruption of the link, in this study the travel speed at the damaged link will be modified to represent the interruption. Such an approach is justified by the work flow in a transportation demand model, where the link length is usually obtained from geographic information system (GIS) databases; travel times are usually estimated by volume-delay functions (BPR-like) to estimate travel time based on travel speed and capacity. Thus, one can easily modify the equations so that travel speed and capacity are modified by multipliers representing the damage and the gradual recovery during reconstruction stages. The weight w_{ij} can represent the number of daily trips as well as the economic value or the tonnage transported between nodes i and j.

The time required for reconstruction can be incorporated into the analysis using a stage-based reconstruction process, which is usually true for most real-life construction events. Therefore, network capacity is reestablished in cumulative discrete steps as selected damaged roads are rebuilt in each stage. At the end of each stage, those roads are back to predisaster conditions (in terms of capacity and free-flow travel time); thus, the network capacity increases step by step, until all roads have been selected and rebuilt. Hence, considering T as the total time required to rebuild all damaged roads, bridges, and tunnels, defined by a finite number of reconstruction stages, each stage with an expected duration d_t (usually in days or weeks), the total impact in accessibility I_f is defined by Equation 9.4.

$$I_f = \sum_{t \in T} \{A_k(t) - A_k(t_0)\} \times d_t \tag{9.4}$$

where I_f is the total impact in accessibility, including reconstruction period.

As a convention, it is suggested that t_0 represents the period before the disaster, t_1 represents the period immediately after the disaster and before reconstruction begins, and the subsequent periods $t_2, t_3, \ldots, t_n$ represent the reconstruction stages. Figure 9.5 illustrates the sudden reduction in network capacity due to the disaster and its gradual increase due to stage-based reconstruction.

The evaluation of the risk related to the disruption of critical elements in the infrastructure permits one to evaluate the risks

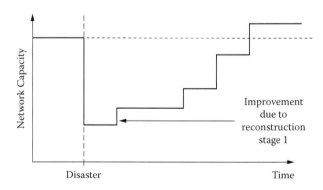

FIGURE 9.5 Stage-based reconstruction.

involving the passenger or freight transport operation and the location of production facilities in relation to the location of suppliers and clients.

9.3.1 Facility Location

Facility location is a very well studied branch of logistics planning. Diverse methodologies have been developed for the selection of optimized locations, as described in Chapter 4. Two aspects of natural disasters are to be considered when deciding a facility location: the vulnerability of the main facility location to natural disasters and the location of backup facilities, which are intended to ensure business continuity in case of disruption of the main facility due to disasters.

Ratick, Meacham, and Aoyama (2008) mention several disasters encompassing large areas where backup facilities turned out to be very beneficial. Perhaps the most commonly mentioned methodology to the determination of a backup facility is the location covering models. Special attention should be placed on the definition of emergency facilities location due to the state of the network during emergencies.

9.4 CONCLUSION

Acquiring resilience against disasters involves a strong perception that disasters will happen and that developing the ability to analyze the potential impacts on social and private infrastructure is of major importance. For private companies, knowledge about their own vulnerabilities and adequate management mind-set are of paramount importance to the development of efficient mitigation and response plans. Understanding the relation between the company's production process and lifeline—especially transportation infrastructure, which is shared with the rest of society—is crucial for designing decision support strategies to cope with the impact of large-scale disasters.

REFERENCES

BSI (Bundesamt für Sicherheit in der Informationstechnik) (2004). Kritische Infrastrukturen in Staat und Gesellschaft. Available from http://www.bsi.bund.de/fachthem/kritis/index.htm

CRED (Center for Research on the Epidemiology of Disasters) (2007). *Annual disaster statistical review: Numbers and trends 2006.* Brussels, Belgium: University of Louvain.

Chang, L. W. (2001). A study on the scheduling of the emergency rehabilitation operations after a major disaster. Master thesis, Tamkang University.

Chang, S. E. (2000). Transportation performance, disaster vulnerability, and long-term effects of earthquakes. Working draft.

Chang, S. E., and Nojima, N. (2001). Measuring post-disaster transportation system performance: The 1995 Kobe earthquake in comparative perspective. *Transportation Research Part A* 35:475–494.

Chen, W. Y., and Tzeng, G. H. (1999). A fuzzy multiobjective model for reconstructing post-earthquake road network by genetic algorithm. *International Journal of Fuzzy Systems* 1 (2): 85–95.

Eusgeld, I., Kroger, W., Sansavini, G., Schlapfer M., and Zio, E. (2009). The role of network theory and object-oriented modeling within a framework for the vulnerability analysis of critical infrastructures. *Reliable Engineering and System Safety* 94:954–963.

Feng, C. M., and Wen, C. C. (2005). A fuzzy bi-level and multi-objective model to control traffic flow into the disaster area post earthquake. *Journal of the Eastern Asia Society for Transportation Studies* 6:4253–4268.

Finn, W. D. L., Ventura, C. E., and Schuster, N. D. (1995). Ground motions during the 1994 Northridge earthquake. *Journal of Canada Civil Engineering* 22:300–315.

Gibb, F., and Buchanan, S. (2006). A framework for business continuity management. *International Journal of Information Management* 26:128–141.

IRGC (International Risk Government Council) (2006). White paper no. 1: Risk governance—Towards an integrative approach. Geneva: International Risk Government Council.

Kaplan, S., and Garrick, B. J. (1981). On the quantitative definition of risk. *Risk Analysis* 1:11–27.

Leaning, J. (2008). Disasters and emergency planning. Harvard School of Public Health, pp. 204–215.

Mote, J., Copeland, C., and Fischer, J. (2003). Critical infrastructures: What makes infrastructures critical? In *Report for Congress*. Congressional Research Service, Library of Congress, Washington, DC.

Perrow, C. (1984). *Normal accidents: Living with high-risk technologies.* New York: Basic Books, Harper Collins Publishers.

Ratick, S., Meacham, B., and Aoyama, Y. (2008). Locating backup facilities to enhance supply chain disaster resilience. *Growth and Change* 39 (4): 642–666.

Rose, A., Benavides, J., Chang, S. E., Szczesniak, P., and Lim, D. (1997). The regional economic impact of an earthquake: Direct and indirect effects of electricity lifeline disruptions. *Journal of Regional Science* 37:437–458.

Tsai, C., and Chen, C. (2009). An earthquake disaster management mechanism based on risk assessment for the tourism industry: A case study from the island of Taiwan. *Tourism Management* 31:470–481.

Wu, H. C. (1997). The research of post-earthquake highway network repairing construction schedule. Master thesis, Institute of Traffic and Transportation National Chiao Tung University, Taiwan.

Yamada, Y., Iemura, H., Noda, S., and Izuno, K. (1986). Evaluation of the optimum restoration process for transportation systems after seismic disaster. *Japan Society of Civil Engineers* 368/I-5:366–362.

Yeh, C. H. (1999). Study on loss assessment of earthquake disaster and scenario simulation methods. *Science Development* 27:260–268.

CHAPTER 10

Application of ICT and ITS

Takayoshi Yokota and Dai Tamagawa

CONTENTS

10.1 Introduction 245
10.2 Development Area of ICT and ITS 246
    10.2.1 Traffic Information System 246
        10.2.1.1 Vehicle Information and Communication System 246
        10.2.1.2 Probe Car System 247
    10.2.2 Electronic Toll Collection Systems 249
        10.2.2.1 What Is ETC? 249
        10.2.2.2 Data Collection Using ETC 250
        10.2.2.3 Multipurpose Uses of ETC 250
    10.2.3 Fleet Management System 250
    10.2.4 Assistance for Safe Driving 251
10.3 Effects of ITS 252
10.4 Next Generation Road Services with ICT and ITS 252
Reference 253

10.1 INTRODUCTION

ICT (intelligent communication systems) and ITS (intelligent transport systems) are key factors in urban transport and logistics. Recently, Internet technology and wireless communication technology have advanced greatly, and many things which used to be desires have been realized. This chapter describes the current status of the ICT and ITS and their potential contributions to urban transport and logistics.

10.2 DEVELOPMENT AREA OF ICT AND ITS

10.2.1 Traffic Information System

10.2.1.1 Vehicle Information and Communication System

The traffic information system is one of the effective systems that have been developed in many countries. For instance, in Japan, an advanced traveler information system named VICS, which is an acronym for "vehicle information and communication system," was developed in the late 1990s under the initiative of the national police agency; Ministry of Land, Infrastructure, Transport, and Tourism; and Ministry of Public Management, Home Affairs, Post and Telecommunications, as well as private sectors, and has been operating since 1996. VICS gathers traffic information from many authorities with variety of sensors. In the following, each sensor is described:

- Most basic sensors are ultrasonic sensors, which detect traffic volume and occupancies' estimated velocity, and some type of them can estimate rough vehicle type, whether it is large or small. Many of the ultrasonic sensors have long been installed for the purpose of traffic signal control and coordination and used also as the sensor for the traffic information system.
- Infrared sensors utilize infrared rays instead of the ultrasound devoted to the VICS, and the functionality is almost identical to the ultrasonic sensors. In addition, this infrared sensor has a communication function to vehicles equipped with the VICS three-media ready-car navigation system. Each car navigation system measures travel time between adjacent infrared sensors and transmits the travel time.
- Travel time sensors using license plate recognition, which is rather expensive, give highly accurate travel time data with vehicle type. Because of their cost, not many have been installed.

Since 1996, VICS has been deployed quickly throughout Japan. The goal of VICS is to provide real-time traffic information to drivers in order to give them decision-making evidence.

The raw traffic sensor data are processed so that the car navigation system can interpret them. Each raw traffic datum is aggregated onto a standard digital road map database called the VICS link. Travel time is estimated from spot velocity data by ultrasonic sensor and infrared sensor, which are optimally mapped onto VICS links. On the other hand, if travel time itself has been acquired, it is divided and

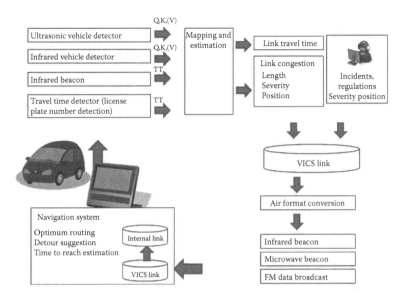

FIGURE 10.1 Outline of traffic information system (VICS).

mapped onto the VICS link. Figure 10.1 shows the priority in which the sensor data are used. In Europe and in the United States, so-called RDS-TMC (radio data service, traffic message channel), transport protocol exports group (TPEG), and or the like has been deployed. Major differences are the kinds of media on which the information is broadcast.

10.2.1.2 Probe Car System

The probe car system became popular in the 2000s with some early work at BMW's XFCD (extended floating car data system) on which the focus was mainly on safety issues. In the United States, a traffic information service based on nearly a million probe cars together with conventional roadside sensor data has been in service (INRIX). In Asia, Japan, China, and Singapore, probe car systems have been tested and some of them are in operation. In the probe car system, every probe car acts as a sensor that moves with the vehicle. For instance, if the vehicle is in congestion, the velocity of probe data will decrease, or position updates of the vehicle may change very little. If we gather these probe data, we can know the whole picture of the traffic situation of the road network. As Figure 10.2 shows, the system architecture is relatively flexible compared to the VICS.

Several auto manufacturers have already started their own services using probe cars. Among many applications, traffic information service is the most effective application of probe car systems. The processing of

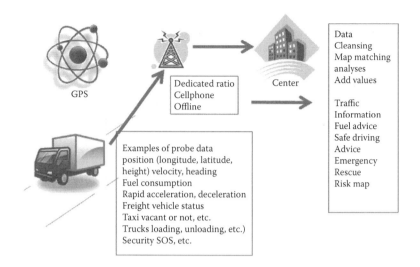

FIGURE 10.2 Outline of probe car system.

the probe car data to achieve traffic information is summarized in the following:

a. Get probe raw data.
b. Eliminate outlier.
c. Map matching to digital road map to get link travel time.
d. From (c), the velocity profile along the link is also obtained.
e. Convert link-based traffic information to suitable presentation form.
f. Provide the traffic information in a suitable protocol.

The coverage of the probe car is an important issue. Fushiki and Yokota reported a new formulation for it:

$$\beta = (1 - \exp(-Q't))^2 + Q't \exp(-2Q't) \qquad (10.1)$$

where
β is spatial coverage
Q' is probe-car traffic volume
t is update interval

This equation gives the estimate for the coverage of the traffic information with a given probe car traffic volume and desired information update interval (see Figures 10.3 and 10.4).

Application of ICT and ITS   249

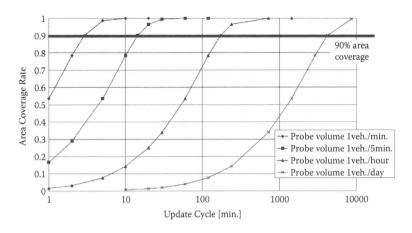

FIGURE 10.3 Relationship of coverage of probe car system.

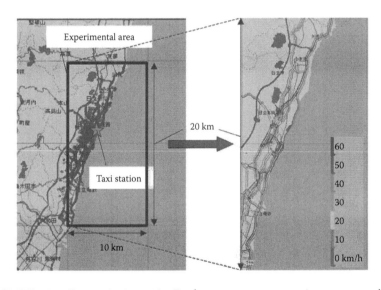

FIGURE 10.4 (See color insert.) Probe car system experiment example in Hitachi, Japan.

10.2.2 Electronic Toll Collection Systems

10.2.2.1 What Is ETC?

In Japan, many traffic jams occurred at motorways, and one of the main reasons was that drivers had to stop at the toll gate to pay the toll. Based on such background, an ETC (electronic toll collection) system was developed. At ETC systems in Japan, the toll is paid by wireless communications between an in-vehicle device mounted onto the vehicle and an antenna at the toll gate, and cars can pass through without stopping. In ETC systems in Japan, an IC card that identifies the individual attribute of users is needed, and the card must be inserted in the in-vehicle device when using ETC. Now, the ETC system is used in many countries around the world, but the structures vary among countries or even areas in the same country.

ETC toll gates can handle several times more cars than non-ETC toll gates; thus, traffic jams would be reduced as ETC came into wider use. Now, in Japan, it is possible to use ETC at almost all motorways and the rate of using ETC at the toll gate has reached over 80% (as of July 2009).

The toll can be changed flexibly by using the ETC system, and various types of tolls can be realized such as changing by time or by route. Also, as an ETC toll gate can handle more cars, less area is needed when placing the tolling point. Thus, new tolling points can easily be added by adopting ETC.

10.2.2.2 Data Collection Using ETC

Various information, such as place or time that each vehicle enters and leaves the motorway, is collected and recorded by ETC systems in Japan. Thus, analysis of origin and destination (OD) volume and vehicles' speed in the motorway and so on are undertaken by using ETC data. When ETC systems had not been developed and used, a questionnaire was used to determine OD volume and so on. However, now that many vehicles use ETC, OD volume can be estimated with accuracy. Also, daily behavior can be researched because the data of each vehicle are recorded in ETC. However, as private information is included, sufficient attention must be paid when using these ETC data.

10.2.2.3 Multipurpose Uses of ETC

Recently, another different use of ETC has been studied in Japan. For example, there are several parking lots in which the ETC system is used for charging the parking fee. Also, as an experiment for a limited time, an ETC system has been used for charging passage fare of cars when boarding ferries. However, the number of available places is still limited and these uses have not spread widely for now.

10.2.3 Fleet Management System

A fleet management system is similar to the probe car system. Usually, a fleet management system consists of the following functions:

  a. Vehicle monitoring system: This system keeps track of the location and status of each vehicle. By this function, the company

can monitor whether the vehicle is on schedule or not. This is almost the same functionality as that of the probe cars; however, the probe data are mainly focused on the status of the vehicle rather than the traffic situation.
b. Safe driving support system.
c. Examining the digital tacograph data or probe data will give us a precious data of each driver's driving skill and risk. It is widely known that the speed variance has a correlation with the risk of accidents.
d. Vehicle routing and scheduling.
e. There are many ways to achieve optimum or suboptimum routes and schedules. Another chapter in this book will describe these in detail.
f. Establish a communication environment between shipper and carrier.

Most of the leading surface transport companies are equipped with these fleet management systems, which reduce the operation cost.

10.2.4 Assistance for Safe Driving

Safe driving is a key issue in ITS and logistics. There are many projects concerning this issue. They are roughly divided into two categories: (1) roadside-equipment-based systems, and (2) vehicle-based systems. The former usually consists of sensors, communication devices to vehicles, or message signs. In Japan, a system called DSSS (driving safety support systems) that is mainly focused on reducing accidents at intersections has been tried. It uses an infrared beacon as a sensor and communication device. Roadside sensors detect pedestrians, bicycles, and so forth in a vehicle's blind spot and communicate the information to drivers. A more highway-oriented system called *smartway* has been tried by using DSRC (dedicated short range communication) devices.

Some of the features of smartway are the following:

a. Vehicle behavior acquisition in an icy, cold area
b. Support for prevention of collisions with forward obstacles
c. Corner overshooting prevention
d. Merging support
e. Speeding vehicle warning

These have been tested in the field in five regions in Japan since 2008 and are supposed to be deployed gradually from 2010.

Vehicle-based systems assist safe driving ranges in collision avoidance systems based on radar or lasers and cameras to speed an adaptation system based on navigation systems. The collision avoidance system keeps track of the distance between one vehicle and another; when the distance falls below some lower limit, it will warn the driver or avoid acceleration in order to avoid the collision.

10.3 EFFECTS OF ITS

Development of ITS is effective for reducing several negative impacts. For example, according to the popularity of using ETC in Japan, the capacity to handle cars at toll gates grew and traffic jams at toll gates have been reduced as a result. The reduction of traffic jams not only releases stresses of drivers but also reduces exhaust gases such as NO_x and CO_2. Reduction of NO_x is favorable for residents who live in roadside areas. Reduction of CO_2 is effective for stopping global warming. It was reported that reduction of CO_2 was estimated at about 140,000 t of CO_2 per year when use of ETC reached 60% (ITS Promotion Office et al. 2007).

10.4 NEXT GENERATION ROAD SERVICES WITH ICT AND ITS

After the beginning of the twenty-first century, global warming has become a serious issue in the world. Several auto manufacturers have announced environmentally friendly hybrid vehicles and electrical vehicles. However, it is not sure that large vehicles for logistics can also be replaced with such technologies. Even though the large vehicles are not going to change greatly, advanced ICT and ITS will contribute to reducing energy consumption and CO_2 emission.

For instance, more efficient routing and scheduling should contribute. To achieve this goal, the variation or reliability of the road network should be measured more precisely. In this regard, the roadside-sensor-based traffic information system is not enough; combining a probe system's and electronic toll collection system's data must be required. If we know the good points and bad points of the road network more than we do now, the system will produce more optimum solutions. If e-commerce and ITS/ICT combine, all the necessary information is gathered and transferred among buyers, producers, retailers, and carriers. And cooperative transport will be realized in order to minimize the cost and maximize the benefits to society, companies, and individuals.

REFERENCE

ITS Promotion Office; Road Administration Division; Ministry of Land, Infrastructure, Transport and Tourism; ITS Planning and Promotion Office; Japan Institute of Construction Engineering (2007). ITS—A collection of effectiveness case studies: 2007–2008.

CHAPTER 11

Future Perspectives on Urban Freight Transport

Eiichi Taniguchi and Russell G. Thompson

CONTENTS

11.1 Challenges of Urban Freight Transport	255
11.2 The City Logistics Rationale	256
11.3 City Logistics and Urbanization Trends	256
11.4 The Future of Cities	257
11.5 Logistics Providers and Urbanization	257
11.6 City Logistics and Efficiency	258
11.7 Disasters and Urban Freight Transport	259
Reference	259

11.1 CHALLENGES OF URBAN FREIGHT TRANSPORT

There are growing concerns about environmental issues relating to urban freight transport due to rising levels of congestion caused by the increasing number of passenger and freight vehicles operating in urban areas. The concentration of people in urban areas in many countries of the world has accelerated this tendency.

Discussion relating to the sustainable development of urban areas as well as mobility of goods has led to more attention to coordinate both traffic and logistics problems. There is an urgent need for more efficient and effective freight transport systems in terms of logistics costs, with full consideration of environmental issues including noise, air pollution, vibration, and visual intrusion. To address these challenges, the concept of city logistics was developed for establishing more efficient and environmentally friendly urban logistics systems. City logistics has been defined as "the process for totally optimizing the logistics and transport activities by private companies with the support of advanced information systems in urban areas considering the traffic environment, its congestion, safety and energy savings within the framework of a market economy" (Taniguchi et al. 2001). This definition describes the conceptual ideas of city logistics.

However, in order to establish efficient and environmentally friendly urban logistics systems through the process of city logistics, visions for city logistics must be created. First of all, it is necessary to set targets for the activities that can be achieved using city logistics. In this context three targets are presented: mobility, sustainability, and livability. Mobility is the central element for ensuring smooth and reliable traffic flow including freight traffic. Sustainability is also important for making cities more environmentally friendly to reduce greenhouse gas emissions as well as to decrease impacts on the local environment. Most recently, livability has become more essential with the increased number of elderly residents in cities.

11.2 THE CITY LOGISTICS RATIONALE

City logistics is needed to establish mobile, sustainable, and livable cities. Urban freight issues are very complicated due to the differing objectives and behavior of the major stakeholders (shippers, freight carriers, administrators, and residents or consumers) involved in urban freight transport. Each stakeholder has different goals and interests based on his or her own role.

The city logistics approach can help solve complicated and difficult problems of urban freight transport using the technical innovations of intelligent transport systems (ITS) and information and communication technology (ICT), as well as encouraging a behavioral change of private companies based on corporate social responsibility (CSR). For example, the development of an urban distribution center for cooperative freight transport among competitive carriers could be achieved through the process of public–private coordination. A successful example of a cooperative freight transport system can be observed in the Motomachi shopping district in Yokohama, Japan. In this case, the city logistics approach was helpful for implementing effective policy measures and maintaining them for mobile, sustainable, and livable freight transport in urban areas.

11.3 CITY LOGISTICS AND URBANIZATION TRENDS

In the next decade, we will face increasing urbanization in which more people will be concentrated in urban areas for a better quality of life. As well, with an aging population, elderly people with mobility problems will have a growing need for home delivery services for their daily commodities. Currently, about half of the population of the world lives in urban areas, and this is estimated by the United Nations to increase to over 60% by 2030.

Firstly, if more people use passenger cars for traveling in urban areas with limited road network capacity, higher levels of road congestion as well as the more negative impacts on the environment will be generated. To alleviate congestion and environmental nuisance, city logistics measures will be required. City logistics solutions will help solve these problems by making more efficient use of freight vehicles and economic/regulatory measures. Since cities must rely heavily on trucks and vans for urban distribution and there are no practical public transport systems available for freight, city logistics will provide a powerful approach to tackle these issues.

Secondly, elderly persons living alone who cannot move by themselves will require delivery of food and other daily commodities as well as medical and daytime care services. For society at large, this type of delivery service using e-commerce will be important for the well-being of cities from the viewpoint of welfare and security. To reduce the costs of individual delivery services to homes, city logistics schemes will need to be introduced.

11.4 THE FUTURE OF CITIES

Some cities in developed countries in the future may face a decrease in population but have a higher proportion of elderly people due to lower birth rates and better medical care. As a result, cities may grow smaller instead of larger, but they need to be more livable and healthy. This trend in some cities is supported by the strategic policy measures of "smart decline." The concept of smart decline includes more efficient use of limited land and transport systems in an aging society.

The use of public transport, including buses, tramways, and bicycles, is encouraged rather than use of private cars, since public transport can promote people's health as well as stimulate the community, lower energy consumption, and provide a better environment. In this context, city logistics can also play an important role. For example, access to shopping streets in a city center by passenger cars and trucks should be limited by regulation. Passenger car traffic can be shifted to buses or tramways, but truck traffic should be well managed using city logistics measures, such as off-hour delivery or cooperative freight transport systems.

11.5 LOGISTICS PROVIDERS AND URBANIZATION

Logistics providers can participate in the coordination of public–private partnerships. These entities are very important for sharing the knowledge and experience of stakeholders, as well as planning and implementing challenging policy measures for livable communities. Logistics providers are often reluctant to join public–private partnerships, since they are competing and do not want to share their information relating to customers and costs of production, transport, and inventory. However, collaboration with shippers, residents, and administrators associated with urban goods distribution allows logistics providers to enhance their position in urban society and be accepted as good partners for making cities more livable and healthy. It is basically helpful for them to continue their business with less impact for residents and to provide better levels of service to their customers.

11.6 CITY LOGISTICS AND EFFICIENCY

In order to facilitate city logistics and more efficient logistics for cities, the attitude of stakeholders who are involved in urban freight transport should change. They need to recognize the importance of working together from the initial stages of the urban planning process. Discussing issues and finding approaches and solutions for urban freight transport problems are beneficial for all players. As well, implementing city logistics measures and evaluating the results are also essential for gaining feedback to improve policy measures.

Although the deregulation of the logistics sector is good for stimulating competition, regulation should be implemented and well enforced in urban areas for freight vehicles and freight traffic to ensure smooth traffic flow and better environments. ITS and ICT can be used for enforcing traffic regulations and management schemes—for example, congestion charging and road pricing for heavy trucks using video cameras. Sometimes subsidies are required to help shippers and freight carriers start new initiatives for reducing the negative impacts on the environment, since a huge initial investment is often needed to start new schemes. If new schemes such as urban consolidation centers and intermodal freight terminals are financially difficult for shippers and carriers, subsidies from the national government or municipality should be provided.

City logistics will have a vital role in enhancing the mobility, sustainability, and livability of cities in the future. This will require a collaborative framework for all urban freight stakeholders, including a change in the mind-set of all players including logistics providers. City logistics will also become increasingly important in an aging society to create healthier and more secure communities.

11.7 DISASTERS AND URBAN FREIGHT TRANSPORT

In the case of disasters, urban freight transport is also important for providing emergency relief supplies to people who are affected by disasters. Special planning and management is required to cope with these issues. "Humanitarian logistics" is the term used to describe logistics operations in emergency cases. Humanitarian logistics aims at minimizing the suffering of affected people by disasters. A shortage of water, food, blankets, fuel, and other daily products was encountered in the Haiti earthquake (2010), the Tohoku earthquake (2011), and superstorm Sandy (2012). The shortage of emergency goods required was mainly caused by the damage to the transport and communication infrastructure as well as the lack of planning, management, and testing of emergency goods distribution systems. When preparing emergency goods distribution systems for disasters in the future, lessons learned from previous disasters should be well applied for improved operations.

REFERENCE

Taniguchi, E., Thompson, R. G., Yamada, T., and van Duin, R. (2001). *City logistics: Network modeling and intelligent transport systems.* Oxford, England: Pergamon.

Index

A
Accelerometers, 63–65, 67
Actigraph accelerometer, 64
Active transport, 57, 58
Administrators as stakeholders in city logistics, 2
Air quality, 16
  Asian cities, issues in, 34, 40–41
  particulate matter (PM), 34
Asian traffic management systems. *See* Traffic management systems, Asia
Automated Resource for Chemical Hazard Incident Evaluation (ARCHIE), 87

B
Benefit-cost analysis, evaluating for walking and cycling tracks, 70–72
Bilevel programming, 170, 178
Body mass index (BMI), 17
Bridge weigh-in-motion (BWIM), 148, 150–151
  accuracy, 157–158
  Bangkok, Thailand, case study, 158–163
  crack response of reinforced concrete slab, estimation by, 157
  live loads, 155
  second members, estimation of responses by, 155–156
  strain responses, main girders, 151–156
Burden of disease (BoD), 16

Business continuity management (BCM), 237. *See also* Business continuity plans (BCPs)
Business continuity plans (BCPs). *See also* Business continuity management (BCM)
  accessibility, evaluations of, 239–240
  economic disruptions, 236
  Kobe port example, 236–237
  overview, 236
  risk evaluation, 240–241
  transportation disruptions, 237–238
Business to Consumer (B2C) distribution, 12

C
Cancer, 54
Capacitated vehicle routing problem (CVRP), 94
Center for Research on the Epidemiology of Disasters (CRED), 226
Centrifugal acceleration, 146–147
Chance-constrained models, 207. *See also* Stochastic vehicle routing problem (SVRP)
City logistics
  conceptual ideas of, 256
  definition of, 2
  efficiency, 258
  future cities, 257
  health issues; *see* Health issues, city logistics

human security engineering; *see*
   Human security engineering
impact of, 258
land use planning for, 6
livability, goal of; *see* Livability
mobility, goal of; *see* Mobility
optimization models; *see*
   Optimization models, city
   logistics
overview, 2
pillars of, 3
providers of, 257–258
rationale for, 256
resilience; *see* Resilience, in city
   logistics
stakeholders, 258
sustainability, goal of; *see*
   Sustainability
urbanization trends, relationship
   between, 256–257
vision of, 3, 4, 5
Conflict analyses, traffic. *See also*
   Traffic conflict technique
   (TCT)
centrifugal acceleration, 146–147
data used, 144–145
lane departure dangers, 145–146
Meihan National Highway case
   example, 142
Nakahata curve case example, 143
outlines, use of, 141–142
overview, 139
possibility index for collision
   with urgent deceleration
   (PICUD), 140–141
time to collision, 139–140
Corporate Social Responsibility
   (CSR), 16
Corporate social responsibility (CSR),
   256
Crash statistics
classifications, 126–127
data sources, 127
exposure, 127–128

D
Dedicated short range communication
   (DSRC) devices, 251–252

Degree of dynamism, 194–195, 199
Diabetes mellitus, 16, 54
Disability adjusted life year (DALY),
   16, 55
Disaster recovery planning, 235
Disasters, defining, 225–226. *See
   also* Man-made disasters;
   Natural disasters
Discrete network design problem
modeling framework, 170–171
MPEC-based, 178–179
overview, 168–169
Distinguished plans, 219
Driving safety support systems
   (DSSS), 251
Dynamic customers, 193
Dynamic stochastic hedging heuristic
   (DSHH), 219–220
Dynamic travel time, 197–199

E
Ecology, urban. *See* Urban ecology
Electronic toll collection systems,
   249–250
Emissions
health, impact on, 15–16
levels, Asian cities, 40–41
reductions through technology,
   40–41, 252
Energy balance modeling, 70

F
Facility locations, 241
Fleet management
   systems, 250–251
Freight carriers. *See also* Urban
   freight transport
efficiency issues, 5
environmental issues, 5
goals of, 2
Freight transport network (FTN)
discrete network design problem;
   see Discrete network design
   problem
overview, 167–168
Freight transport, urban. *See* Urban
   freight transport

G

Genetic algorithm (GA), 94–95
Genetic local search, 179–181
Geocoding, 63
Global position systems (GPSs), 67–68, 69

H

Haiti earthquake, 259
Hazardous materials facilities, 95
Hazardous materials transport, 20–21
  accident types related to, 79, 81
  accidents, number of, 81–82, 84
  causes of accidents related to, 82, 84
  constraints, 79
  defining, 77–78
  economic loss from accidents, 82
  increase in, 79
  multiple parties, 84–85
  number of shipments, 79
  risk management; *see* Risk management, hazmat
  routing; *see* Routing, hazmat
Hazardous road locations (HRLs)
  analysis, in-office, 129
  counter-measures, 131–132
  effectiveness of road safety activities, evaluating, 130–131
  identifying, 128–129
  prevention of crashes, 130
  programs aimed at reducing, 129
Health issues, city logistics
  accelerometers, use of; *see* Accelerometers
  active transport; *see* Active transport
  air quality; *see* Air quality
  cancers, 54
  disability adjusted life year (DALY); *see* Disability adjusted life year (DALY)
  diseases related to, 54, 55
  geocoding; *see* Geocoding
  health and transport, relationship between, 54
  journey planning, 60–61
  multicriteria methods of evaluation, 71–72
  obesity; *see* Obesity
  overview, 15–16
  Personal Health, 16
  physical activity, 17–18, 56–57, 68–69
  sedentary life styles, 17, 54–55, 56
  self-change, 60
  self-reported measurement regarding, 62–63
  walkability; *see* Walkability of environments
  years of life lost (YLL); *see* Years of life lost (YLL)
Highway Capacity Manual, 107
Human security engineering
  man-made disasters; *see* Man-made disasters
  natural disasters; *see* Natural disasters
  overview, 18–19
  postdisaster demands, 19
  resilience, building in; *see* Resilience, in city logistics
Hurricane Katrina, 7
Hybrid vehicles, 252
Hyogoken-Nanbu earthquake, 236
Hypertension, 54

I

Information communication technology (ICT), 5, 20. *See also specific technology*
  importance to urban transport and logistics, 245
  next-generation road services, 252
  traffic information systems, 246
Infrared sensors, 246
Infrastructure, defining, 229
Infrastructure, natural disaster impacts. *See under* Natural disasters
Intelligent transport systems (ITS), 5, 101. *See also specific systems*

contributions to safety, potential, 138–139
effects of, 252
enforcement, role of, 137
importance, 137, 245
information collection, 139
next-generation road services, 252
overview, 137
traffic information systems, 246
International physical activity questionnaire (IPAQ), 62

J
Journey planning, 60–61

K
Katrina, Hurricane, 7

L
Land use mix, 59
Land use planning, 6
  Asian cities, 42, 43
  integrated, 43
List-based threshold accepting (LBTA) algorithm, 94
Livability, goal of, in city logistics, 3, 4
Load limits, 150
Logistics planning, 241

M
Man-made disasters. *See also specific disasters and events; Natural disasters*
  economic disruptions, 236
  human security engineering, as part of, 18
  infrastructure security, need for, 19
  supply chain strategies, 19–20
  types of, 19
Master delivery schedule (MDS), 12
Mathematical program with equilibrium constraints (MPEC), 170, 171, 174, 178
Megalopolis
  effects of, 44–45
  emergence of, in Asia, 44, 46–47
  features of, 46
  implications of, 46
Meihan National Highway case example, 142
Melbourne, Australia, bushfires, 7
Miner's law, 148, 150
Mixed-traffic flow
  road safety issues, 117–120
  traffic management systems, Asia, 105–107
Mobility, defining, 3–4
Motomachi shopping district, 256
Multiobjective evolutionary algorithm (MOEA), 215
Multiobjective optimization problem (MOP), 13. *See also Optimization models, city logistics*
Multiple plans approaches (MPAs), 219

N
Nakahata curve case example, 143
National Public Health Partnership, 55
Natural disasters. *See also specific disasters and events*
  classifying, 226–227
  economic disruptions, 236
  human security engineering, as part of, 18
  impact analyses, 227–228
  infrastructure, impact on, 227, 228–229
  reconstruction impacts, 231–232
  recovery planning process, 235
  risk estimation, qualitative, 234–235
  risk management, 232–233
  transportation disruptions, 237–238
  transportation operations, impact on, 229–230
Niigataken Chuetsuoki earthquake, 8
North American Emergency Response Guidebook, 87

O
Obesity, 17, 55
Olympics (Sydney), simulation used in planning, 12

Optimization models, city logistics
  multiagent simulation, 14–15
  multiobjective optimization,
    12–14, 20
  robustness, 10–11
  simulation, 12
  stochastic programming, 11–12
  usage, 10
OPTIPATH, 93
Osteoporosis, 54

P

Pareto optimal solution, 13
Passenger car units (PCUs), 107,
  108–109
Physical activity guidelines, 17
Pipeline and Hazardous Materials
  Safety Administration's
  (PHMSA), 87
Piracy, Somalia, 8
Planimate simulation software, 12
Poisson distribution, 210
Possibility index for collision
  with urgent deceleration
  (PICUD), 140–141
Probabilistic vehicle routing and
  scheduling problem with
  time windows (P-VRPTW)
  application, 216–217
  evolution of, 215–216
  SVRP, *versus*, 215
  time travel distribution, 216
Probe car systems, 247–248

Q

Q-learning process, 15
Quantitative risk measurement,
  233–234

R

Recourse model, 207, 209–210, 214
Residents, urban, as stakeholders in
  city logistics, 2
Resilience, in city logistics
  human security engineering, as
    part of, 18
  pillar of city logistics, as, 3, 4
Resilience, transport systems, 228

Risk
  assessment, hazmat; *see* Risk
    assessment, hazmat
  defining, 10
  governance, 10
  management of, hazmat; *see* Risk
    management, hazmat
  planning for, in urban freight
    transport, 10
  qualitative risk estimation,
    234–235
  quantitative risk measurement,
    233–234
Risk assessment, hazmat. *See also*
    Risk management, hazmat
  aims of, 88
  overview, 87–88
  probability calculations, 88–89,
    88–90
  steps in, 88
Risk management, disaster, 232–233
Risk management, hazmat. *See also*
    Risk assessment, hazmat
  compliance, 86
  exemptions, 86
  importance, 85
  mitigation, 86–87
  overview, 85
  regulations, 85–86
  self-evaluation, 857
Road safety
  audits; *see* Road safety audits
  crash statistics; *see* Crash statistics
  human component, 124–125
  mixed-traffic flow, issues related
    to, 117–120
  roads component, 124, 125–126,
    128–129, 130; *see also*
    Hazardous road locations
    (HRLs)
  vehicle component, 124–125
Road safety audits
  overview, 132–133
  process of, 133–134
  road design analysis, 134
  safe road environments, 134
Rolling horizon, 196
Routing, hazmat

global, 91–92
government issues, 91–92
local, 91
problems related to, 90, 91
schedule planning, 91
shortest path models, 92–93, 94, 95
vehicle routing and scheduling with time windows (VRPTW) models; see Vehicle routing and scheduling with time windows (VRPTW) models
vehicle routing problems models, 94–95

S

Safe road designs, 135
Safety, structural. See Structural safety
Sandy, superstorm, 259
Sarin attacks, Tokyo subway, 7
SC-T-SNE case example, 174–178
Sedentary lifestyles. See under Health issues, city logistics
September 11 attacks, 7–8, 19
Shippers
 goals of, 2
Sichuan earthquake, 7
Signalized intersections, 107
Singapore, transport and logistics in, 43–44
Smartway, 251
Solomon's benchmark instances, 199, 218–219
Somalia, piracy attacks, 8
Stochastic and dynamic vehicle routing problem with time windows (SDVRPTW)
 instances, 218–219
 overview, 217–218
 solution approaches, 219–220
 waiting strategies, 218
Stochastic programming, 11–12
Stochastic user equilibrium (SUE), 169
Stochastic vehicle routing problem (SVRP)
 chance constrained model, 207–208
 exact approaches, 211–213
 heuristic approaches, 214–215
 instances, 210
 overview, 206–207
Structural safety
 bridge weigh-in-motion (BWIM); see Bridge weigh-in-motion (BWIM)
 fatigue damage, 148, 150
 heavy vehicles, impacts of, 148, 150
 miner's law, 148, 150
 overloaded vehicles in Asian countries, 148
 overview, 147–148
Sumatra earthquake, 7
Supply chain management, 167
 long-term strategy of, 168
Supply chain network (SCN), 167
 decision makers of, 172
 efficiency, 174
 influence of, 168
 planning, 170
Sustainability
 city development in Asia, 49
 goal of, in city logistics, 3
 importance, 4
 measurement of, 2
Sydney Olympics, simulation used in planning, 12

T

Tabu searches, 204, 214
TABUSTOCH, 214
Tohuku earthquake, 259
Traffic conflict technique (TCT), 139
Traffic Engineering Handbook, 107
Traffic management systems, Asia, 102
 intersection designs, 110–112
 mixed-traffic flow, 105–107, 117–120
 motorcycle use, 105, 106, 111, 112, 117
 optimal approach design, 115–116
 passenger car units (PCUs), 107, 108–109
 signalized intersections, 107
 traffic capacity estimation, 112–115
 traffic levels, 102–105

Traffic safety, 21
  accident trends, 135–136
  importance, 137
  safety studies, 136–137
Transport and logistics, Asian cities
  air pollution issues, 34
  alternative freight transport and distribution modes, 48–49
  enhancing of transport, 47
  environmental impacts, 40–41
  freight restrictions, 43
  freight traffic, 35, 37–38
  land use planning, 42, 43, 47
  megalopolis, rise of; see Megalopolis
  overview, 31–32, 38
  passenger transport, 42–43
  population, 32–33
  public transportation, 33–34
  social impacts, 39–40
  sustainability, 49
  transportation logistics, 36–37
  travel safety, 38–39
  urbanization, consequences of, 32–33
Transport network equilibrium models, 171–173
  multiclass multimodal case example, 173–174
Tsunami, Japan, 190

U

Urban ecology, 58
Urban freight transport
  challenges of, 255–256
  charges for, 7
  disasters, importance during, 259
  importance of considering in urban planning, 5–6
  planning, 6
  risks, 7–10
  stakeholders, 8
  subsidies for, 6–7
  sustainability, relationship between, 5–6
Urban logistics. *See* City logistics
User equilibrium (UE), 169, 171

V

Value of the Stochastic Solution (VSS), 11
Vehicle routing and scheduling with time windows (-dynamic) (VRPTW-D) models, 9, 15
  customer, 193
  degree of dynamism, 194–196
  diversion issue, 196
  dynamic travel time, 197–199
  exact optimization, 201–203
  heuristics, 203–206
  overview, 193
  rejection policy, 197
  solution approaches, 200–201
  test instances, 199–200
  waiting policy, 196–197
Vehicle routing and scheduling with time windows (-probabilistic) (VRPTW-P) models, 9
Vehicle routing and scheduling with time windows (VRPTW) models, 9, 94, 95
  constraints, 191
  overview, 190–191
  soft time windows, 192
Vehicle routing problem with soft time windows (VRPSTW) model, 11
Vehicle trajectory data (VTD), 144–145
VICS system (Japan), 246–247
Victorian Integrated Survey of Travel and Activities (VISTA), 62
VISSIM (Verkehr in Städten— Simulations), 12

W

Walkability of environments
  civil engineering, role of in creating, 59–60
  index measuring, 59
  neighborhood factors, 58, 59
  overview, 58

Y

Years of life lost (YLL), 16, 55